Introduction to Communication Systems Simulation

For a listing of recent titles in the *Artech House Mobile Communications Library,* turn to the back of this book.

Introduction to Communication Systems Simulation

Maurice Schiff

ARTECH HOUSE
BOSTON | LONDON
artechhouse.com

Library of Congress Cataloging-in-Publication Data
A catalog record of this book is available from the U.S. Library of Congress.

British Library Cataloguing in Publication Data
Schiff, Maurice
   Introduction to communication systems simulation.
   1. Telecommunications systems—Computer simulation
   I. Title
   621.3'82'0113

   ISBN-10   1-59693-002-0

Cover design by Yekaterina Ratner

© 2006 ARTECH HOUSE, INC.
685 Canton Street
Norwood, MA 02062

International Standard Book Number: 1-59693-002-0

10 9 8 7 6 5 4 3 2 1

To my wife Sharon, for the many hours
I devoted to this book that I couldn't devote to her.

Contents

Preface

The genesis for this book was my involvement with the development of the SystemView (now SystemVue) simulation program at Elanix, Inc. Over several years of development, technical support, and seminars, several issues kept recurring. One common question was, "How do you simulate (such and such)?" The second set of issues was based on modern communication systems, and why particular developers did what they did. This book is an attempt to gather these issues into a single comprehensive source.

Chapter 1 presents a discussion of the basic elements of a communication system. It serves as a reference for subsequent chapters by briefly describing the various components of a communication system and their role in the system.

Chapter 2 develops the theory of linear time invariant (LTI) systems, which is the foundation of communication theory. The basic concepts of the filter impulse response and convolutions are presented. From there we consider the workhorses of LTI systems, namely the Fourier and Laplace transforms. We end with the simple development of the fast Fourier transform (FFT), which has revolutionized signal processing.

Chapter 3 deals with the concept of sampling. As digital processors become faster, more and more of the system processing is performed in this domain. A thorough understanding of the Nyquist sampling theorem and other issues is vital to the successful implementation of these systems.

Chapter 4 provides the fundamentals of filters, as they are the most common element of a communication system. We start with the concept of filter phase and group delay via a simple two-tone input example. From there we separate the discussion into the two common classes of filter finite impulse response (FIR), and infinite impulse response (IIR).

Chapter 5 concentrates on the concept of digital detection. Most modern communication systems use digital formats as opposed to analog (AM, FM). The fundamental results of optimum digital detection are derived along with several equivalent processing architectures.

Chapter 6 is concerned with the various methods of conveying information in a digital format. Physically, the transmitted wave is a sine function. One can convey information only by modulating the amplitude, frequency, phase, or combinations thereof this basic signal. This chapter also provides a discussion of spread spectrum modulation concepts including both frequency hopping and direct sequence.

Chapter 7 is the complement of Chapter 6 in that it describes techniques for demodulating the transmitted signal at the receiver. The basic architecture of in-phase and quadrature processing is detailed. In addition, the methods for recovering frequency, phase, and data timing (synchronization) are considered.

Chapter 8 deals with the important concept of baseband pulse shaping. The radio frequency spectrum is an economic quantity that the United States government auctions to service providers. The goal of the provider is to provide as many data channels as possible within this spectrum. This goal is commonly achieved by limiting the signal spectra by using baseband processing such as raised and root raised cosine filters.

Chapter 9 presents the heart of digital communication performance analysis: the bit error rate (BER) calculation. Here we detail methods for simulating BER calculations. Two issues are of concern. The first is how to match the timing of the input data stream with the recovered one. The system under test will have various delays via encoders, filters, and so on, so the output data will be delayed in time with respect to the input. The second issue is how to calibrate the signal-to-noise ratio (SNR) for proper interpretation of the results.

Chapter 10 describes what happens to the signal between the transmitter and receiver. This is called the channel. Fading is the most important channel. The signal can bounce off of various objects and present to the receiver several versions of itself modified in amplitude, frequency, phase, and time of arrival. The effect of this is the well-known fading phenomena. This chapter describes various forms of the fading, as well as methods to mitigate their effects.

Chapter 11 deals with the real world of nonlinear amplifiers. The nature of these nonlinearities are described and modeled. Both standard RF amplifiers and satellite-based traveling wave tube amplifiers are discussed.

Chapter 12 concerns the simulation of communication systems using baseband concepts. Unless there are nonlinear elements in the RF section, one can mathematically eliminate the carrier operations from the simulation. This is important since simulating the carrier requires a very high system sample rate to adequately describe the carrier. By contrast, the information content on the carrier may have a information bandwidth several orders of magnitude smaller. The baseband simulation allows accurate simulation while only sampling at a rate sufficient to represent the information source. This presents orders of magnitude savings in the simulation time.

Chapter 13 concludes the book with a look at the emerging technology of ultra-wideband (UWB) systems. The two competing technologies, direct sequence (DS) and orthogonal frequency division multiplexing (OFDM) are presented.

As with any lecture course, an attendant laboratory can be used to emphasize and further enhance the material through direct demonstrations. To this end we have included a disc containing example files using the simulation software SystemVue. The reader is urged to run these files, and vary the various system parameters as permitted. It is hoped that these examples will greatly enhance the readers understanding of the text material.

CHAPTER 1

Elements of a Communication System

In this first chapter we introduce the basic components of a communication system. The system diagram is shown in Figure 1.1.

We have emphasized the major elements of the system as the transmitter, channel, and receiver. Each of these elements has several subcomponents. We shall work our way through this diagram starting with the information source, and ending with the recovered information. The ideas and issues pertinent to each sub-block will be described.

1.1   The Transmitter

The task of the transmitter is to take an information source and after several steps of processing, produce a signal at some carrier frequency over the air.

1.1.1   The Data Source

Since we are dealing with digital communications, information will, regardless of the source, eventually be converted into bits. However, the origins of the information may be either digital or analog.

Types of sources that might be considered digital include a text message (e.g., Stop), a numeric value (e.g., 78°F), or a tuning command to some object (e.g., a missile). As an example, the Global Positioning Satellite (GPS) system has a 50-bps data message that provides the user with system time information and the ephemeris data of the transmitting satellite.

The most common types of analog sources are voice and pictures. These sources are sampled, according to the rules of Chapter 1, and the results are converted to a bit pattern.

1.1.2   Source Encoding

In many cases, the data is not independent from one sample to the next; there is some correlation or redundancy. Mathematical techniques called compression algorithms can be applied to the data. This compression reduces the number of bits that must be transmitted, or equivalent, the system bit rate. This is highly desirable since bit rate translates into required system bandwidth, which is usually a premium economic quantity. At the receiving end, these same algorithms recover the original message without loss of information.

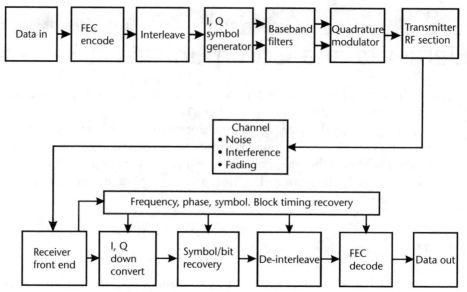

Figure 1.1   Block diagram of a communication system.

One simple algorithm is called *run length encoding*. In this scheme a numeric sequence (5, 9, 6, ...) tells us that the information is five 0s, followed by nine 1s, followed by six 0s, and so on. The .zip application on personal computers (PCs) is a well-known compression system.

1.1.2.1   Audio or Voice Compression

The goal is always to reduce the transmitted bit rate of a system, thus reducing the required over-the-air bandwidth, which, in turn, allows for more users to be accommodated in a prescribed bandwidth. This is a very important issue for the wireless community.

Another area is antijam systems. In these systems the input bit rate is greatly expanded by codes such as pseudo random (PN) or Gold codes. The processing gain, or antijam effectiveness, is the ratio of this high-speed data rate to the input data rate.

The human voice generally occupies a bandwidth from 300 Hz to 3 kHz. In Chapter 3 we show that to sample such a source, a minimum rate of 6 Kbps is required. The digital phone system uses 8 Kbps with 8 bits of amplitude, producing a 64-Kbps data stream. For wireless systems, this rate is much too high. However, the information from one 8-Kbps sample to the next is generally not independent, so some form of compression algorithm is used.

Delta Modulator

The delta modulator is a simple device. The output at time $t + 1$ is a "1" if the signal at $t + 1$ is greater than the signal at t; otherwise, the output bit is a "0". Figure 1.2 shows the basic block diagram of the delta modulator. The demodulator is simply an integrator, and this is shown in the figure as well.

Figure 1.3 shows the overlay of the input and output signal. Figure 1.4 shows the resulting binary data stream derived from the input signal.

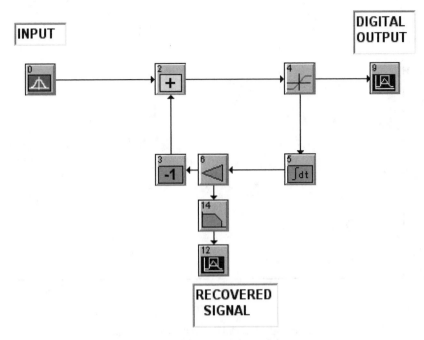

Figure 1.2   Block diagram of a delta modulator and demodulator.

Sigma Delta Converter

Sigma delta converters are very powerful. They are commercially available from a variety of vendors. Figure 1.5 shows a simple version of this converter. The advantage of this technique lies in the quantization noise (i.e., the difference between the original signal and the quantized version).

Figure 1.6 shows the spectrum of the quantization noise for both the simple analog to digital (A/D), and the sigma delta quantizer. Note in Figure 1.6 the low frequency role-off of the quantization noise for this technique, giving it improved performance over the ordinary quantizer.

Pulse Code Modulation

With pulse code modulation (PCM), the analog time sample is quantized into 2^n levels, which can then be represented by n bits. If Rs is the symbol sample rate, then the resulting bit rate will be $Rb = nRs$. This is accomplished by using an analog limiter called a compression expander (COMPANDER). The COMPANDER is inserted between the analog source and the sampler. There are two commonly used standards.

1. $\mu = 255$ Law: This version is common in the United States and has the transfer function

$$z = z_{max} \, \text{sgn}(x) \ln\left[1 + \mu |x| / x_{max}\right] / \ln[1 + \mu]$$

$$\text{sgn}(x) = 1 \text{ for } x \geq 0$$

$$= -1 \text{ for } x < 0$$

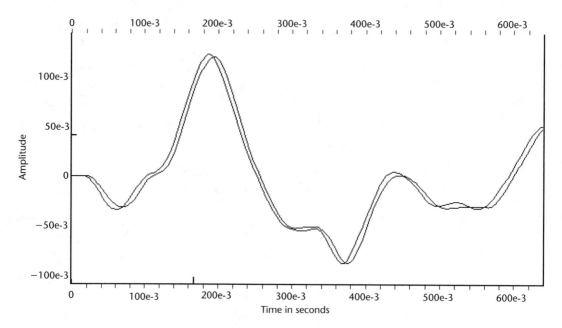

Figure 1.3   Overlay of the input and demodulated output of the delta modulator.

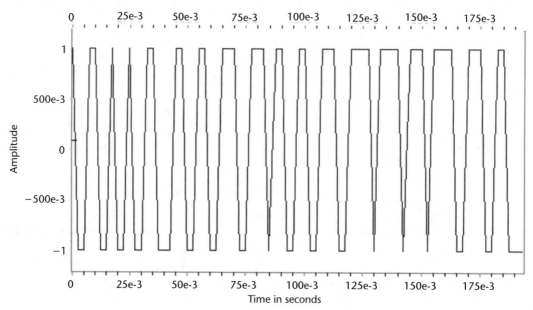

Figure 1.4   Binary data stream derived from the input signal of Figure 1.3.

2. $A = 87.6$ Law: This version is used in Europe has the transfer function

$$z = z_{max}\, \mathrm{sgn}(x) A\left(|x|/x_{max}\right)/[1 + \ln A] \qquad \text{if } 0 < |x|/x_{max} \le 1/A$$
$$= z_{max}\, \mathrm{sgn}(x)\left[1 + \ln\left(A|x|/x_{max}\right)/[1 + \ln A]\right] \quad \text{if } 1/A < |x|/x_{max} < 1$$

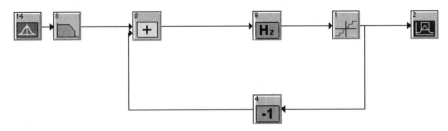

Figure 1.5   Sigma delta converter block diagram.

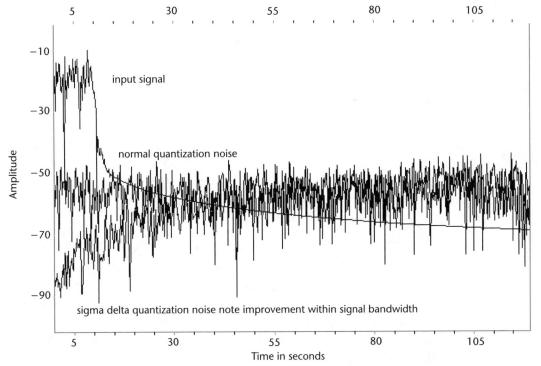

Figure 1.6   Comparison of the quantization noise between the delta modulator, and the sigma delta converter. Note the role of the quantization noise at low frequency for the sigma delta converter. This increases the SNR for a specified number of quantization bits.

The phone system uses a system with samples at 8 Kbps, and 8 bits/sample, giving a transmitted data rate of 64 Kbps. Figure 1.7 is an overlay of these transfer functions.

Vocoders
Vocoders are conversion algorithms working on voice input. The human voice is quite redundant, and these compression algorithms can produce output rater far less than that of the 64-Kbps PCM system just mentioned. Vocoders are complicated algorithms that work on blocks of data usually 5 to 20 ms long. The basic idea is to model the physiology of the voice tract and extract features such as pitch.

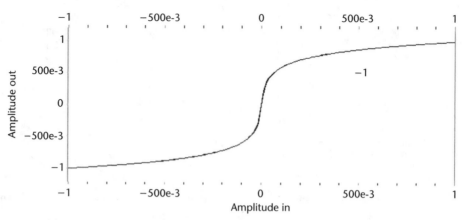

Figure 1.7 Overlay of the $\mu = 255$ and A = 87.6 compander transfer functions. Note that despite their different mathematical description, they are virtually indistinguishable.

1.1.3 Forward Error Correction

Forward error correction (FEC) coding is a system to improve the system BER performance. The basic concept is a process that takes in k bits and encodes them into $n > k$ bits. The ratio $r = k/n$ is called the rate of the code. This action will increase the required system bandwidth, but it is well worth it in practice. It is therefore desirable to keep r as close to unity as possible. Commonly used values are $r = 1/2$ and $r = 2/3$. There are several types of FEC codes.

Block FEC
As the name implies, this encoder takes in block of k bits, and performs the encoding process to produce a block of n bits. The rate, r, of the code is $r = K/n$. Powerful mathematical systems have been used to provide efficient decoding algorithms. The number of transmitted errors that the decoder can correct is usually denoted by t. Thus, the 23-bit triple error correcting Golay code has $(n, k, t) = (23, 12, 3)$. It is a perfect code in that it will correct any combination of three or fewer errors, but no combination greater than three errors. Other common block codes include the Base Chadhur Hocquenghem (BCH) and Reed-Solomon codes.

The BER performance of a block code can be approximated by the formula

$$P_e \propto e^{-r(tt1)E_b/N_0}$$

Where E_b/N_0 is the energy per bit, as described in Chapter 5. Without the coding the error rate is

$$P_e \propto e^{-E_b/N_0}$$

What is happening is a trade-off. The energy per encoded bit is less than the original energy per bit since we put out more encoded bits in the same amount of time.

The power of the code can be increased by taking larger blocks of n and k, such that the code rate remains nearly constant, but the correction capability t increases. From the above two relations, we see that for a code to be effective it must have the relation $r(t+1) > 1$. For the Golay code just mentioned we have, $r(t+1) = [12/23]*4$ $= 2.09$

Convolutional FEC

The convolutional FEC encoder works more like a tapped delay line filter, as shown in Figure 1.8.

The input data is sequentially clocked through the series of K shift registers, as shown. Attached to these registers are sets of logic, two in this case, that produce one output bit each per input clock cycle. In this case, the diagram gives an $r = 1/2$ rate code. The number of registers, K, is called the constraint length of the code: the larger the K, the more powerful the code. The decoding algorithm that revolutionized the use of convolutional codes was derived by Viterbi, and the algorithm bears his name. One problem with this algorithm is that the computational load grows exponentially with the constraint length K. Thus, the question of the existence of "good" short K codes was an issue. This question was answered positively by a paper written by Heller and Jacobs.

One major advantage of the Viterbi algorithm is that it can operate on soft decisions. In a hard decision decoder, a decision 1 or 0 (hard) is made on the individual bits *before* the decoder. It can be shown that this leads to a performance loss of about 2 dB. The Viterbi algorithm can be modified to work on a quantized bit, usually 3 bits or eight levels, which recover most of this loss. Such a modification is extremely difficult for a block decoder.

When simulating FEC codes, remember that the sample rate changes across the FEC coder and back again across the decoder. For a rate 1/2 code, the output rate is twice the input. Care must be taken in the algorithm code to handle this situation.

1.1.4   Interleaving

If the only effect of the channel is to add white noise, the errors caused will be statistically random. However, it possible for the channel to become suddenly very bad, due to fading and other effects, thus obliterating a whole block of data. One solution to this problem is to interleave the data. The idea is to take a successive block of

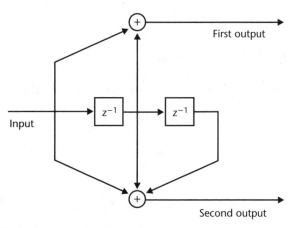

Figure 1.8   Rate 1/2, $K = 3$ convolutional encoder.

k input bits and separate them with bits from successive blocks in such a way that the out block contains only 1 bit per input block. The effects of the fade are mitigated by the deinterleaver at the receiver, producing one error per block that the FEC can generally handle.

The most common interleaver is the block interleaver. In this method an array of say $n \times k$ bits is established. The incoming bits are stored row by row. When the array is full, the bits are read out column by column. A second type is the convolutional interleaver. Table 1.1 shows a 3×3 array.

The data is read in by rows (1, 2, 3, 4, 5, 6, 7, 8, 9), and read out by columns (1, 4, 7, 2, 5, 8, 3, 6, 9).

1.1.5 Baseband Filtering

Baseband filtering is used to reduce the bandwidth of the transmitted signal. The objective is to perform this filtering without introducing intersymbol interference (ISI). In Chapter 8 we take up this operation in detail.

1.1.6 Modulation

A modulated signal, $s(t)$, on a carrier f_0 can always be written in the form

$$s(t) = A(t)\sin\left[2\pi f_0 t + \varphi(t)\right]$$

where we note that frequency modulation is another form of phase modulation since the two are related by a derivative, $f(t) = d\varphi(t)/dt$. A convenient alternative to this representation is

$$s(t) = I(t)\cos(2\pi f_0 t) + Q(t)\sin(2\pi f_0 t)$$
$$I(t) = A(t)\sin\left[\varphi(t)\right]$$
$$Q(t) = A(t)\cos\left[\varphi(t)\right]$$

In Chapter 6 we will describe several standard formats such as PSK, QPSK, and QAM.

1.2 The Transmission Channel

The transmission channel is the medium between the transmit antenna, and the receive antenna.

Table 1.1 Interleaver Array

1	2	3
4	5	6
7	8	9

The data is read in by rows and read out by columns.

1.2.1   Additive White Gaussian Noise

Additive white Gaussian noise (AWGN) is the most fundamental impairment. The term "white" implies that all frequencies are of equal strength, and "Gaussian" describes the amplitude distribution. Note that these are *independent* concepts. You can have one without the other.

The receive antenna essentially sees thermal radiation with a nominal 290°K temperature. The exception to this is when the antenna is looking out to space, as a satellite receive station antennal would be. In this case, deep space is considered a black body with a 4°K temperature. Another deep space noise source is the so-called galactic noise. A space-borne antenna, which sees part of the Earth and part of deep space, will have an equivalent temperature at some intermediate temperature value.

1.2.2   Interference

The radio spectrum is crowded. All sorts of users are transmitting all sorts of signals. Some of the interferers might be directly in band. The classic case of an in-band interferer is an intentional jammer trying to deny a military link. In other cases, strong signals in adjacent frequencies can spill energy into the desired band. Most receiver components such as amplifiers are nonlinear. A strong signal in any band can then produce harmonics just about anywhere.

1.2.3   Fading

The signal from the transmitter can bounce off objects in the immediate area of the transmission link, as shown in Figure 1.9. These reflected signals can arrive at the receiver and combine with the direct signal.

In the simplest concept, consider one reflected path that arrives at the receiver with the same amplitude, but 180° out of phase with the direct signal. The result is complete signal cancellation.

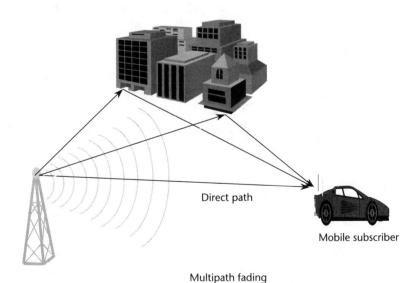

Direct path

Mobile subscriber

Multipath fading

Figure 1.9   General fading modes.

Fading is generally divided into several categories. A flat fade is one where all frequency components of the signal are affected equally (i.e., the spectrum "breathes" up and down as a whole). A selective fade is one where only specific frequencies are affected, so the received spectrum appears to have notches. A fast fade is one where the channel dynamics are faster than the information rate.

In many systems it is this interference that limits the system performance, not the AWGN.

1.3 Receiver

A receiver is orders of magnitude more complicated to design in the physical world or to simulate than the transmitter. This is because the receiver must recover all sorts of system parameters before a successful demodulation can be performed.

1.3.1 Frequency Offset

Even if the transmitter and receiver are not moving with respect to each other, there are frequency offset problems. What the transmitter clock says is 100 MHz is not necessarily what the receiver clock says is 100 MHz. Sometimes this effect is negligible. The more accurate the clocks, the better, but they become more expensive as well.

Even in a perfect clock world, there can be motion between the transmitter and receiver. Mobile wireless is a perfect example. This motion produces a Doppler shift of the transmitted frequency by an amount $f_d = (v/c)f_0$, where f_0 is the transmitted carrier frequency, v is the component of velocity toward the receiver, and c is the speed of light.

In the signal intercept area, an observer sees a "blob" of energy on a spectrum analyzer. He then places makers on this energy to queue an local oscillator (LO) down conversion. However, this frequency estimate is approximate, and in fact can be quite large.

In Chapter 7, we will describe several algorithms for obtaining these offsets.

1.3.2 Phase Offset

Many modulations convey the information in the phase of the transmitted signal. For QPSK, the typical coding of 2 bits to a phase is shown in Table 1.2.

It should be noted that the phase of the receiver oscillators are not synched to the phase of the transmit oscillators, even if their frequencies are identical. Thus,

Table 1.2 QPSK Symbol Representation

Bits	Phase
00	0
01	$\pi/2$
10	π
11	$3\pi/2$

received values can be shifted by some unknown value α. If $\alpha = \pi/2$, a complete decoding failure would occur.

Tracking loops such as the Costas loops have been designed to eliminate this offset. But even these generally have an ambiguity problem. The loops cannot differentiate the case when the received phase has been rotated into another possible phase as mentioned above.

The solution is to differentially encode the phase. The decoder recovers the information by subtracting two successive phase estimates, thus eliminating this ambiguity. In the wireless IS-95 and other code division multiple access (CDMA) systems, the base station actually transmits an unmodulated data signal that the receiver can use to lock to and recover the absolute phase reference.

1.3.3   Timing Offset

The next problem is system timing. The fundamental timing unit is the encoded bit or symbol. The receiver needs to know when to sample the raw demodulated waveform to recover the information in an optimum manner. The last stage of a demodulator is usually a matched filter that is sampled once per symbol, but at the correct time to maximize the SNR.

If the system has FEC, then the timing that tells the FEC decoder where a data block begins must be established. This is called word or frame sync.

1.3.4   Data Recovery

At this point we have recovered the basic data symbols. What remains to be done is to recover the original data bits or analog (voice) content. The procedure is to perform the inverse operation for each of the steps encountered in the transmitter—performed in reverse order, of course. Thus, as required, we deinterleave, FEC decode, and convert via the source compression algorithm back to an analog signal.

1.4   Conclusion

In this chapter, we presented the basic elements of a communication system, from the information source (in) to the recovered information (out). We described the various steps taken in the transmitter, through the channel, and by the receiver to recover the data. The receiver is much more complicated due to the various synchronization steps required for proper data recovery.

Selected Bibliography

Heller, J. A., and Jacobs, I. W., "Viterbi Decoding for Satellite and Space Communications," IEEE Trans. Communtechnol., Vol. COM19 No. 5, Oct. 1971, pp. 835–848.

Pahlavan, K., and A. H. Leveseque, *Wireless Information Networks*, New York: John Wiley & Sons, 1995.

Proakis, J. B., *Digital Communications*, New York: McGraw-Hill 1983.

Rappaport, T. S., *Wireless Communications, Principles and Practice*, Upper Saddle River, NJ: Prentice Hall, 2002.

Sklar, B., *Digital Communications, Fundamentals and Applications*, Upper Saddle River, NJ: Prentice Hall, 2001.

Steele, R., *Mobile Radio Communications*, London: Pentech Press Limited, 1992.

CHAPTER 2
Linear Time Invariant Systems

Linear time invariant (LTI) systems are the backbone of much of the analysis of communication systems, especially filtering operations. In this chapter we develop the concept of an LTI filter in the time domain. This immediately leads to the Fourier transform (FT). Starting with the continuous FT, we continue to the digital Fourier transform (DFT), which is used on a computer. The concept of FT windows is introduced here. Finally, we describe the all-important fast Fourier transform (FFT) algorithm via a specific example.

2.1 LTI Systems

Consider a system described by a time function h(t). We observe $h(t)$ by "kicking" it with a unit impulse response $\delta(t)$, as shown in Figure 2.1. The function $h(t)$ is then called the unit impulse response of the system. A system is said to be time invariant if the impulse is delayed by T, and the output is delayed by the same amount $h(t - T)$; that is, the system does not change in time. A time variant system could occur if someone changed the value of some element of your system, such as a resistor or capacitor. In fact, there are capacitors called varactors that can be controlled by an input voltage, which could be any function of time.

Now let the input signal be a series of impulses, a_k, separated by time T. Then the input can be written as

$$x(t) = \sum_k a_k \delta(t - kT)$$

Then the output becomes

$$y(t) = \sum_k a_k h \ etc.(t - kT)$$

Now take $h(t)$ to be

$$h \ etc.(t) = 1 \quad 0 \leq t \leq 1$$
$$= 0 \quad \text{otherwise}$$

and let the input $x(t)$ be 10 unit height pulses, $a_k = 1$, separated by $T = 0.1$ sec. Then, by the above definitions, the value of $y(t)$ at $t = 1$ is

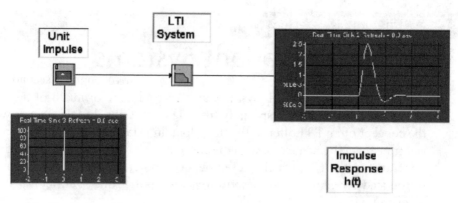

Figure 2.1   Linear system impulse response.

$$y_{10}(1) = \sum_{k=1}^{10} h(t-.1k) = 10$$

Repeat the calculation with $x(t)$ now being 100 unit impulses separated by 0.01 sec. Then at $t = 1$ we have

$$y_{100}(1) = 100$$

Clearly as we continue this process $y_\infty(1)$ becomes infinite. This is not a good thing! However, if we modify the basic in-out relation to

$$y(t) = T\sum_k a_k h(t - kT)$$
$$a_k = x(kT)$$

then $y(1) = 1$ regardless of the spacing of the input impulses. Now we take the limit that $T \rightarrow 0$, and by the rules of calculus we have

$$y(t) = \int_{-\infty}^{\infty} h(t - \tau)x(\tau)d\tau = h(t) * x(t)$$

where the * operator is commonly used as a shortcut notation for the full expression that is called the convolution function. Now if

$$y_1(t) = h(t) * x_1(t)$$
$$y_2(t) = h(t) * x_2(t)$$

then by the basic integral definition we have

$$z(t) = h(t) * \left[ax_1(t) + bx_2(t) \right]$$
$$z(t) = ay_1(t) + by_2(t)$$

This result is the definition of a linear system. From a practical standpoint, in a linear system, having two sine wave functions of different frequencies as an input, only those frequencies will appear at the output.

Now what happens if we take $y(t)$ and pass it through as second filter with impulse response $g(t)$. The result is

$$z(t) = y(t) * g(t) = \left[x(t) * h(t) \right] * g(t)$$

Although the calculation procedure developed here is conceptually simple, it has two major drawbacks: (1) although it is straightforward, the calculations are tedious; and (2) there is the question of how to design an $h(t)$ for some specific purpose.

Enter the Fourier transform.

2.2   Fourier Transform (FT) Theory

The Fourier transform $H(f)$ of a signal $h(t)$ is given by the relation.

$$H(f) = \int_{-\infty}^{\infty} h(t) e^{-2\pi j f t}\, dt$$

where the output variable f is the signal frequency. This is the continuous version of a Fourier series. The inverse transform is

$$h(t) = \int_{-\infty}^{\infty} X(f) e^{2\pi j f t}\, df$$

We now develop one of the most important aspects of FT. Suppose that the function $H(f)$ in the above expression is the product of two other frequency functions $X(f)$, and $Y(f)$. Then the expression for $h(t)$ is

$$h(t) = \int_{-\infty}^{\infty} X(f) Y(f) e^{2\pi j f t}\, df$$

But for some frequency functions $X(f)$ and $Y(f)$ we have

$$X(f) = \int_{-\infty}^{\infty} x(\tau) e^{-2\pi j f \tau}\, d\tau$$

$$Y(f) = \int_{-\infty}^{\infty} Y(\tau) e^{-2\pi j f \hat{\tau}}\, d\hat{\tau}$$

Substituting these expressions gives

$$h(t) = \int_{-\infty}^{\infty}\int_{-\infty}^{\infty}\int_{-\infty}^{\infty} x(\tau)y(\tau)e^{2\pi j(ft-f\tau-f\hat{\tau})}df d\tau d\hat{\tau}$$

$$= \int_{-\infty}^{\infty}\int_{-\infty}^{\infty} x(\hat{\tau})y(\tau)d\tau d\hat{\tau}\left[\int_{-\infty}^{\infty} e^{2\pi jf(t-\tau-\hat{\tau})}df\right]$$

First evaluate the integral in the square brackets by noting the following FT pairs:

$$h(t) = \delta(t - T)$$
$$H(f) = e^{-2\pi jfT}$$

With this relation the expression for $h(t)$ reduces to

$$h(t) = \int_{-\infty}^{\infty}\int_{-\infty}^{\infty} x(\tau)y(\hat{\tau})\delta(t - \tau - \hat{\tau})d\tau d\hat{\tau}$$

The integral with respect to τ is easy, owing to the definition of the delta function. The net result is

$$h(t) = \int_{-\infty}^{\infty} x(t - \hat{\tau})y(\hat{\tau})d\hat{\tau} = x(t) * y(t)$$

This is an important result. We have the same expression for $h(t)$ previously obtained. What this says is to calculate the output of an input function $x(t)$ through a filter $h(t)$; we simply multiply the FT of each signal and perform the inverse FT. By extension, if there are two filters, the output is the inverse FT of the individual elements:

$$z(t) = \int_{-\infty}^{\infty} H(f)G(f)X(f)e^{2\pi jft} dt$$

This operation is shown in Figure 2.2. We now have a relatively simple method of calculation, and, as we shall see shortly, the filters are readily defined in the frequency domain.

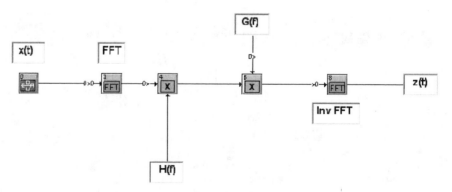

Figure 2.2   Linear system operation on two elements.

Some other useful features of the FT are that:

1. The FT is linear:

$$FT\big[ax(t) + by(t)\big] = aFT\big[x(t)\big] + bFT\big[y(t)\big]$$

2. If the signal $x(t)$ is real, then

$$X(-f) = X^*(f)$$

where the * indicates complex conjugate. Thus it is only necessary to compute the FT for the positive frequencies.

2.3   The Digital Fourier Transform

We now turn to how one simulates an FT on the computer. The founding expression has two problems: (1) it is continuous; and (2) the integral limits are $\pm\infty$. To this end, we modify the FT equation to take these factors into account:

$$H(k\Delta f) = \Delta t \sum_{n=0}^{N-1} h(n\Delta t) e^{-2\pi j k n \Delta f \Delta t}$$

We are evaluating the FT at frequencies that are integer multiples of Δf, and we have sampled the time function at $fs = 1/\Delta t$.

The following relation is basic to the DFT:

$$\Delta f \Delta t = 1/N$$

from which we get the following important results:

$$\Delta f = 1/N\Delta t = 1/T$$

(i.e., the frequency resolution is the inverse of the time length of the segment), and

$$F_{max} = N\Delta F = 1/\Delta t = f_s$$

This is a restatement of the Nyquist criteria.

The eventual choice of these parameters is based on the problem at hand. Clearly $1/\Delta t = f_s$ must be chosen to satisfy the Nyquist criteria for the signal under consideration. Usually, the physics of the problem forces a choice of the frequency resolution ΔF, from which the FFT size N is then calculated.

In standard notation the DFT and inverse DFT expressions become

$$H(k\Delta f) = H_k = \sum_{n=0}^{N-1} h_n e^{-2\pi j k n/N}$$

$$h(n\Delta t) = h_n = \sum_{n=0}^{N-1} H_k e^{2\pi j k n/N} / N$$

It is very instructive to compute the DFT of a sine wave. For convenience we shall use the complex form

$$s(t) = e^{2\pi j f t}$$

Now we let $t = k\Delta t$, $f = m\Delta$ (where m is not necessarily an integer) and define N as before to obtain

$$H_n = \sum_{k=0}^{N-1} e^{2\pi j k(m-n)/N}$$

It turns out that this expression can be evaluated in closed form since it is of the geometric progression form

$$H_n = \sum_{k=0}^{N-1} x^k = \left[1 - x^N\right] / \left[1 - x\right]$$

$$x = e^{2\pi j(m-n)/N}$$

With a little effort we can calculate the magnitude square of H_n. First we substitute the expression for x to obtain

$$H_n = \left[1 - \left(e^{2\pi j p/N}\right)^N\right] / \left[1 - \left(e^{2\pi j p/N}\right)\right]$$

$$= e^{\pi j p} e^{-\pi j p/N} \left[e^{\pi j p} - e^{-\pi j p}\right] / \left[e^{\pi j p/N} - e^{-\pi j p/N}\right]$$

$$p = m - n$$

Finally, using the basic expression for the sine function

$$\sin(x) = \left(e^{jx} - e^{-jx}\right) / 2j$$

we arrive at the final expression

$$\left[H_n\right]^2 = \left[\sin\left[\pi(m-n)\right] / \sin\left[\pi(m-n)/N\right]\right]^2$$

Recall that we evaluate H_n only at frequencies that are integer multiples of Δf, regardless of what f is. So let us start with $m =$ any integer $< N$. Then from the above equation,

$$\left[H_n\right]^2 = N^2 \quad n = m$$
$$= 0 \quad \text{otherwise}$$

This result is plotted in Figure 2.3 with $m = 32$ and a resolution $\Delta f = 1$ Hz. We have the intuitive result that a pure sine wave can exist only at one frequency.

Now let $m = 32.5$. That is, the frequency exactly splits two DFT bin frequencies. The result is shown in Figure 2.4. Notice the major difference. The energy splatters

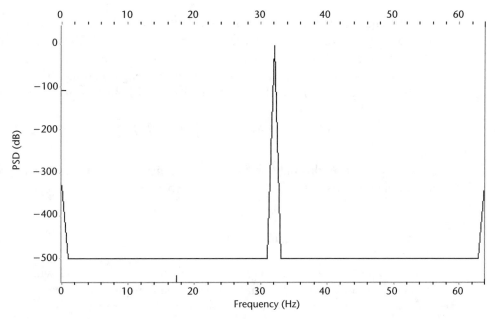

Figure 2.3   DFT of a sine wave, m = integer = 32.

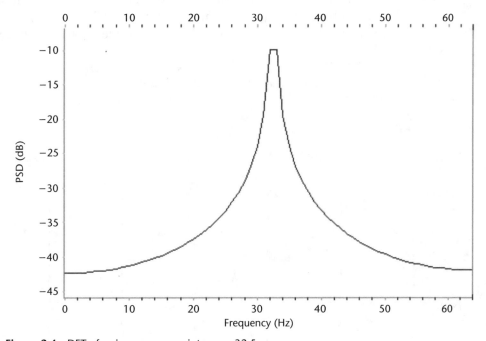

Figure 2.4   DFT of a sine wave, m ≠ integer = 32.5.

all over the frequency domain (it has to go somewhere). Further, the decrease in power as we move away from 32.5 Hz, falls off very slowly as $1/f^2$.

There is another way to arrive at the above results, which may be instructive. The FT is performed only on a segment of the actual signal during a T-second interval. The resulting time domain signal is thus comprised of a sum of sine and cosine waves of frequency that are a multiple of $1/T$. If we consider that we started with a continuous signal over the interval $[0, T]$, we may regard the FFT as the result of taking the *continuous* FT of this signal, and sample it at frequencies k/T. To this end we write the FT of a sine wave of frequency f_0 existing over $[0, T]$ (only the positive part of the magnitude of the FT will be used for clarity without loss of generality),

$$|H(f)| = \left|\sin\left[\{\pi(f - f_0)T\}/\pi(f - f_0)T\right]\right|$$

As will be seen throughout this text, the general mathematical expression $\sin(x)/x$ appears constantly. It is common then to give this relation a special function name:

$$\text{sinc}(x) \equiv \sin(x)/x$$

Figure 2.5 shows what happens if we sample this function at k/T when $f0$ is itself some integer multiple of $1/T$. What we see is the continuous $\text{sinc}(x)$ spectra being samples exactly at the nulls of these spectra, with the one exception where x is equal to zero. Compare this figure to Figure 2.3.

Now repeat the above process when the carrier frequency fo alf way (splits) the FFT sampling; then the picture of Figure 2.6 results. Now we see that the samples are on the main lobes of the $\text{sinc}(x)$. Compare with Figure 2.4.

To see why bin splitting can be troublesome, consider a signal with two sine waves of the form

$$s(t) = .1\sin(2\pi 9t) + \sin(2\pi 11.25t)$$

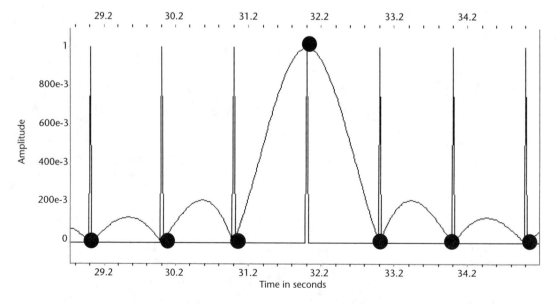

Figure 2.5   The PSD of the sine wave that is on a frequency bin. The samples of the continuous spectra are at the zeros of the $\text{sinc}(x)/x$ function.

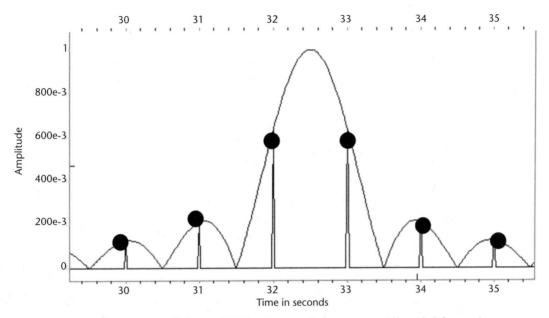

Figure 2.6   PSD of sine wave splitting an FFT bin. The samples now occur at the side lobe maximums.

and perform a DFT on this signal with $\Delta f = 0.5$. The 11.25-Hz signal exactly splits two FFT bins, while the 9-Hz signal is on a bin. The result is shown in Figure 2.7. Notice that the splatter of the stronger 11.25-Hz signal spectra nearly obscures the 10-Hz DFT bin of the weaker signal.

2.4   DFT Windows

Let us return to the transformation from the FT to the DFT. In particular, we have to change the limits of integration from $+\infty$ to $[-T/2, T/2]$. One way of viewing this action is shown in Figure 2.8. This can be interpreted as viewing the entire signal through a window 1 second wide.

The windowed function $s_w(t)$ is then related to the original signal $s(t)$ and the window function $w(t)$ by

$$s_w(t) = s(t)w(t)$$

Now take the FT of the above, giving

$$s_w(f) = S(f) * W(f)$$

Note the reversal of the convolution in the time and frequency domains from the original development of the FT. The window function can be thought of a frequency domain filter. What are the implications of this if the input were a pure sine wave of complex frequency $+fc$? The result is

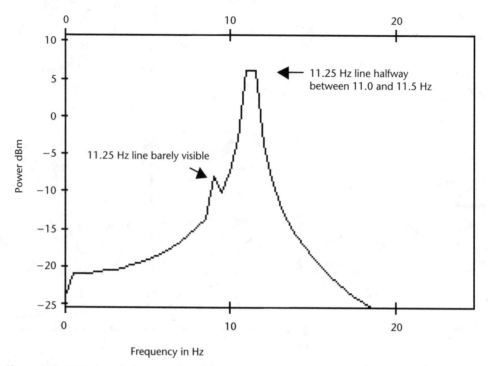

Figure 2.7   DFT of weak sine wave in splatter of a stronger signal.

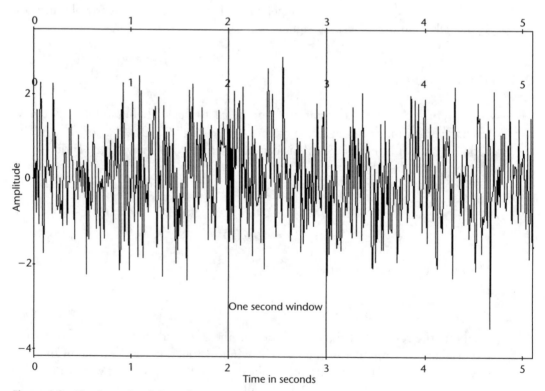

Figure 2.8   Viewing a signal through a 1-second window.

$$w(t) = 1 \quad -T/2 \le t \le T/2 \text{ (rectangle)}$$
$$= 0 \quad \text{otherwise}$$

$$W(f) = \int_{-\infty}^{\infty} w(t)e^{-2\pi j f t}\, dt = \int_{-T/2}^{T/2} e^{-2\pi j f t}\, dt$$

$$= \text{sinc}[\pi f T]$$
$$s(t) = e^{-2\pi f_c t}$$
$$S(f) = \delta(f - f_c)$$
$$S_w(f) = \text{sinc}[\pi(f - f_c)T]$$

So the frequency spectrum of a rectangular windowed pure sine wave has the sinc function as its spectra. As noted before, this function rolls off in frequency very slowly (i.e., it splatters). This result is for a continuous spectrum.

There is extensive literature on windows. For the basic issues of windowing, consider three windows, all defined over $-T/2 < t < T/2$:

1. Rectangular:

$$p(t) = 1$$
$$W^2(f) = \left[\sin(\pi f T)/\pi f^2\right]$$

2. Triangular (Bartlett):

$$p(t) = 1 - |t|/T$$
$$W^2(f) = \left[\sin(\pi f T)/(\pi f T)\right]^4$$

3. Hanning:

$$p(t) = .5\left[1 + \cos(\pi t/T)\right]$$

In the above we normalized all windows to unity amplitude at $f = 0$. Figure 2.9 shows a comparison of the three windows. We observe that as the spectral roll-off increases from (a) to (c), the width of the main lobe increases as well. This is a general result.

It is also instructive to compare the value of each PSD at the maximum of the first side lobe. Table 2.1 compares this value with the spreading of the main lobe for all three windows.

Now we can return to the issue presented in Figure 2.7. In Figure 2.10 we repeat the calculation only using an FFT window. Now the smaller signal is clearly visable, and the strong signal, even though on an FFT bin, now shows the spreading of the spectral width of the sine wave at 9 Hz.

Another use of windows relates to accurate measurement of the PSD of a modulated signal with steep frequency roll-off. A good example is the GMSK modulation of the Global System for Mobile Communications (GSM) wireless system. Figure 2.11 shows a block diagram of this modulation.

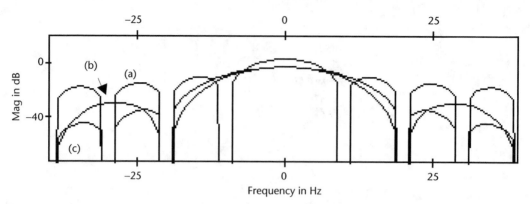

Figure 2.9   Comparison of the PSD of three windows: (a) rectangular, (b) triangular (Bartlett), and (c) Hanning.

Table 2.1   Comparison of Three Windows

	Height of First Side Lobe dB down from f = 0	*Width of Spectral Peak*
Rectangular	13.2	1
Triangular	26.4	1.2
Hanning	42.0	1.4

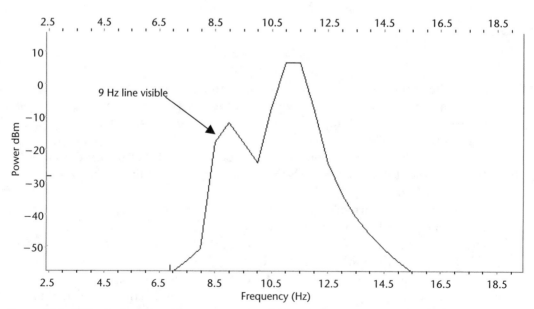

Figure 2.10   PSD of a weak signal near a strong signal. The windowing now cuts down the strong signal splatter making the smaller signal visible.

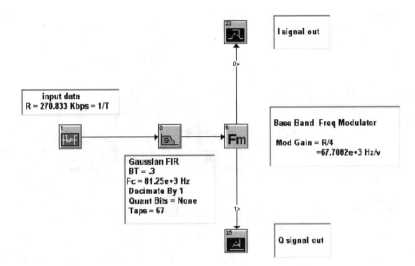

Figure 2.11   Baseband GMSK signal generation. The Gaussian filter is used to severely compress the occupied bandwidth of the signal so it does not splatter into adjacent channels.

Figure 2.12 shows the PSD of this GMSK signal with and without the use of the window. Note that with the window, the skirts properly fall off within the pass band as they should.

2.5   The Fast Fourier Transform

The FFT algorithm has provided an enormous boost to spectral analysis. The desire is to make the DFT length N as large as possible. This decreases the resolution bandwidth Δf (for a fixed sample rate). When detecting a signal and noise, the noise power in a FFT bin is proportional to Δf. This increases the detection sensitivity as N becomes large, and correspondingly Δf becomes small.

The literature is complete with many general derivations of the FFT algorithm. Terms such as "decimation in time," "decimation in frequency," and "butterflies" are common. Our approach is to develop the basic idea of the algorithm by starting with an $N = 4$ DFT, and proceeding from there. Written out, the $N = 4$ DFT looks like the following:

$$H_0 = h_0 W_N^0 + h_1 W_N^0 + h_2 W_N^0 + h_3 W_N^0$$
$$H_1 = h_0 W_N^0 + h_1 W_N^1 + h_2 W_N^2 + h_3 W_N^3$$
$$H_2 = h_0 W_N^0 + h_1 W_N^2 + h_2 W_N^4 + h_3 W_N^6$$
$$H_3 = h_0 W_N^0 + h_1 W_N^3 + h_2 W_N^6 + h_3 W_N^9$$

where

$$W_N^k = e^{-e\pi jk/N}$$

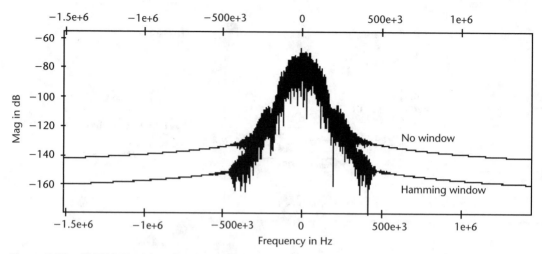

Figure 2.12 GMSK PSD with and without windowing. Notice the Hamming windowed PSD has considerably less out-of-band splatter.

There are $N = 4$ rows, and each row has N multiplications and $N - 1$ additions. For large N, the number of operations, M is

$$M = 2N^2$$

This is not a good result. Doubling N results in a four-fold increase in the computational load. But let us examine the above a little closer. In particular, note the following:

$$W_4^0 = 1 \quad W_4^1 = -j \quad W_4^2 = -1 \quad W_4^3 = j$$
$$W_4^4 = 1 \quad W_4^6 = W_4^4 W_4^2 = -1 \quad W_4^9 = W_4^1 = -j$$

Substituting these results into the basic equations gives

$$H_0 = h_0 + h_1 + h_2 + h_3$$
$$H_1 = h_0 - jh_1 - h_2 + jh_3$$
$$H_2 = h_0 - h_1 + h_2 - h_3$$
$$H_3 = h_0 + jh_1 - h_2 - jh_3$$

now if we define

$$U_\pm = h_0 \pm h_2$$
$$V_\pm = h_1 \pm h_3$$

we arrive at a very simple result

$$H_0 = U_+ + V_+$$
$$H_1 = U_- - jV_-$$
$$H_2 = U_+ - V_+$$
$$H_3 = U_- + jV_-$$

As can be seen, the computational load has been greatly reduced. A general count of the operations using the FFT is

$$M = 2N \log_2 N$$

This is an enormous savings as N becomes large and has made the FFT algorithm the workhorse of the signal processing. As computers have become faster, a common PC can easily do 1 million point FFT in a second or so.

As an aside, we note that the FFT algorithm has been applied to antenna theory. In particular, a Butler matrix feeding n elements will produce n separated beams in an economical way.

2.6 Conclusion

In this chapter we started out with the concept of an LTI system. We showed that the output was the convolution of the input with the impulse response of the system. The continuous Fourier transform was introduced as a mathematical tool that greatly simplifies the calculations, and allows for a convenient design method for developing transfer functions (filters) that serve a desired purpose. Next we developed the DFT, since the continuous FT does not exist in reality, since the limits of integration on the time variable is $\pm\infty$. The concept of an FFT window was introduced and explained as a means of controlling what is generally termed bin splatter, which can hide a small signal. Finally, we developed the FFT algorithm, which revolutionized the use of Fourier theory on computers.

Selected Bibliography

Pahlavan, K., and A. H. Leveseque, *Wireless Information Networks,* New York: John Wiley & Sons, 1995.

Proakis, J. B., *Digital Communications,* New York: McGraw-Hill 1983.

Rappaport, T. S., *Wireless Communications, Principles and Practice,* Upper Saddle River, NJ: Prentice Hall, 2002.

Sklar, B., *Digital Communications, Fundamentals and Applications,* Upper Saddle River, NJ: Prentice Hall, 2001.

Steele, R., *Mobile Radio Communications,* London: Pentech Press Limited, 1992.

CHAPTER 3

Sampling

The single most important concept when simulating communication systems on a computer is that there is no such concept as "continuous." All variables such as time, amplitude, and frequency are stored as numbers in the computer with some sort of quantization. To see why this is true and necessary, consider the following question: How many points are there on a time line segment in the range $0 \le t \le 1$ seconds? The answer, of course, is infinite. One of the strange things in mathematics is that the number of points in the range $0 \le t \le 2$ seconds is the *same* as for the range $0 \le t \le 1$ seconds!

3.1 The Sampling Operation

Given that we must quantize all variables, let us start with time. Time is the independent variable that drives all system simulations. The fundamental question is: What is the smallest time step T that is needed? Usually one specifies sample rate $f_s = 1/T$ in samples/second. If we choose f_s too large, the number of time steps needed to complete the simulation will be very long, slowing the execution time. If we take f_s too small, then the effects of aliasing (to be described below) will seriously affect the simulation output. The answer to this question is the Nyquist Sampling Theorem. The reader is strongly urged to understand this theorem before embarking on a simulation.

Consider any continuous time function $r(t)$. The sampling operation essentially looks at $r(t)$ only at time $t = kT$, with k an integer. To represent this mathematically, we define a sampling function $s(t)$, which has the properties: $s(t) = 1$ for $t = kT$, and $s(t) = 0$ otherwise. Now we obtain the sampled function $r_s(t) = s(t) \cdot r(t)$. Figure 3.1(a–d) shows this operation with $T = 1$ sec.

The sampling function can be written in the form

$$s(t) = \sum_{k=-\infty}^{\infty} \delta(t - kT)$$

where $\delta(x)$ is known as the unit impulse or delta function defined by

$$\delta(x) = 1 \quad x = 0$$
$$= 0 \quad \text{otherwise}$$

$$\int_{-\infty}^{\infty} h(t)\delta(t - a)dt = h(a)$$

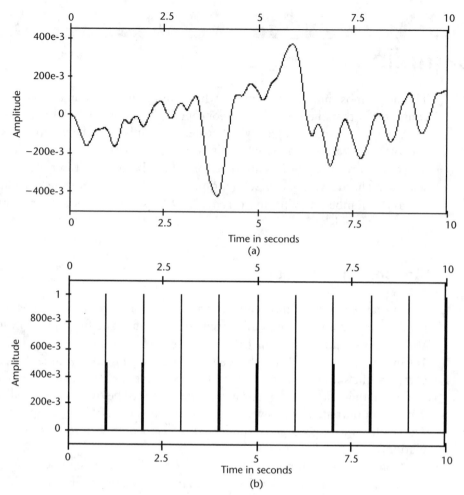

Figure 3.1 (a) Input function $r(t)$; (b) the sampling function; (c) the sampling operation; and(d) the sampled function $r_s(t)$.

where $h(t)$ is any function of t. The true definition of the delta function is the integral relation, while the unit definition is a useful concept.

The expression for the sampled function is now

$$r_s(t) = r(t) \cdot \sum_{K=-\infty}^{\infty} \delta(t - kT)$$

The above is nice, but what implications does it have to the simulation issue? To answer this question we must go to the frequency domain. Remember from Chapter 2 on Fourier analysis, we showed that multiplication in one domain is equivalent to convolution in the other domain. To this end we first need the Fourier transform of the sampling function $s(t)$:

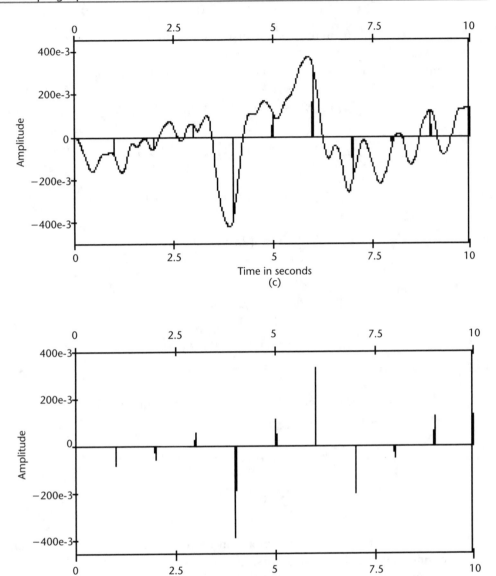

Figure 3.1   (continued).

$$S(f) = \int_{-\infty}^{\infty} s(t) e^{-2\pi jft}\, dt$$

$$= \int_{-\infty}^{\infty} \left[\sum_{k=-\infty}^{\infty} \delta(t - kT) \right] e^{-2\pi jft}\, dt$$

$$= \sum_{k=-\infty}^{\infty} \int_{-\infty}^{\infty} \delta(t - kT) e^{-2\pi jft}\, dt$$

$$= \sum_{k=-\infty}^{\infty} e^{-2\pi jkfT}$$

Even in this form the desired answer is not obvious. To proceed we observe that the sampling function $s(t)$ is periodic with period T; that is, $s(t - kT) = s(t)$ for any integer k. In this case, $s(t)$ can be written as a Fourier series:

$$s(t) = \sum_{k=0}^{\infty}\left[a_k \cos(2\pi kt/T) + b_k \sin(2\pi kt/T)\right]$$

$$a_k = \frac{2}{T}\int_{-1/2\,T}^{1/2\,T} s(t)\cos(2\pi kt/T)dT \quad k \neq 0$$

$$a_k = \frac{1}{T}\int_{-1/2\,T}^{1/2\,T} s(t)\cos(2\pi kt/T)dt \quad k = 0$$

$$b_k = \frac{2}{T}\int_{-1/2\,T}^{1/2\,T} s(t)\sin(2\pi kt/T)dt \quad \text{all } k$$

Using

$$s(t) = 1, \quad t = 0,$$
$$s(t) = 0 \quad \text{otherwise}$$
$$T/2 < t < T/2$$

the expressions for a_k and b_k become: $a_0 = 1$, $a_k = 2$, and $b_k = 0$ for all k. The sampling function now reduces to

$$s(t) = 1 + 2\sum_{k=1}^{\infty}\cos(2\pi kt/T)$$

$$= \sum_{k=-\infty}^{\infty}\delta(t - kT)$$

That is, the sampling function can be considered as an infinite summation of cosine waves, or local oscillators, of frequency k/T.

Now, by remembering the basic relation

$$\left(e^{jx} + e^{-jx}\right)/2 = \cos x$$

the frequency spectra of the sampling function can be rewritten as

$$S(f) = \sum_{k=-\infty}^{\infty}e^{-2\pi jkfT} = 1 + \sum_{k=1}^{\infty}2\cos(2\pi kfT)$$

$$= \sum_{k=-\infty}^{\infty}\delta(f - k/T)$$

$$= \int_{-\infty}^{\infty}\left[1 + \sum_{k=1}^{\infty}2\cos(2\pi kt/T)\right]e^{-2\pi jft}\,dt$$

Thus the Fourier transformation of a sequence of delta functions in the time domain produces a similar sequence of delta functions in the frequency domain.

Now back to the problem at hand, we can write the spectrum $R_s(f)$ of the sampled time function $r_s(t)$ in terms of the spectra of the sampling function $S(f)$ and the original signal $R(f)$, as follows :

$$R_s(f) = S(f) * R(f)$$

$$= \int_{-\infty}^{\infty} S(\alpha)R(f - \alpha)d\alpha$$

$$= \int_{-\infty}^{\infty} \sum_{k=-\infty}^{\infty} \delta(\alpha - k/T)R(f - \alpha)d\alpha$$

$$= \sum_{k=-\infty}^{\infty} R(f - k/T)$$

3.2   The Nyquist Sampling Theorem

The last expression above essentially says that the frequency spectra of the sampled function is comprised of the frequency function of the sampled signal $r(t)$ shifted onto each of the frequency lines $f = k/T$. Figure 3.2(a) shows a signal with a one-sided wide spectrum $B = 2$ Hz, and Figure 3.2(b) shows the resulting spectrum for a signal with a sample rate of 10 Hz. Note that Figure 3.2(b) shows replicas of the original spectrum of Figure 3.2(a), each placed on a multiple of the 10-Hz sample rate.

Remember that we just noticed that the sampling function can be viewed as an infinite number of oscillators separated by $f = k/T$. Thus the sampling function "mixes" the input function onto each of these carriers, giving the same picture in the frequency domain.

Now let us take a closer look at Figure 3.2(b). We see that since $B = 2$ Hz < 10 Hz, the individual spectra are fully separated from one another. Now resample the original signal at a 4-Hz rate. The resulting spectra is shown in Figure 3.2(c).

In Figure 3.2(c) we now see that the individual spectra are just butting up to one another. Clearly if we further reduce the sample rate, these spectra will start to merge into one another. This collision phenomenon is called *aliasing*, and it is a very important concept for understanding how to avoid aliasing problems in your simulation

To avoid aliasing then, we must not violate the well-known Nyquist sampling criteria:

$$B < f_s/2$$

where B is the one-sided bandwidth of the signal being sampled.

This result is very nice, but it assumes that the sampled signal has no spectral content for frequencies $|f| \geq f_2/2$. In the real world, however, *there is no such thing as a band limited signal*! An absolutely band limited signal can only be the Fourier transform of a signal that is *infinite* in time extent. So unless you had the foresight and ability to start your simulation at the dawn of the ages, there is always some aliasing. Ok, what's next then? There is no universal answer to this question. The

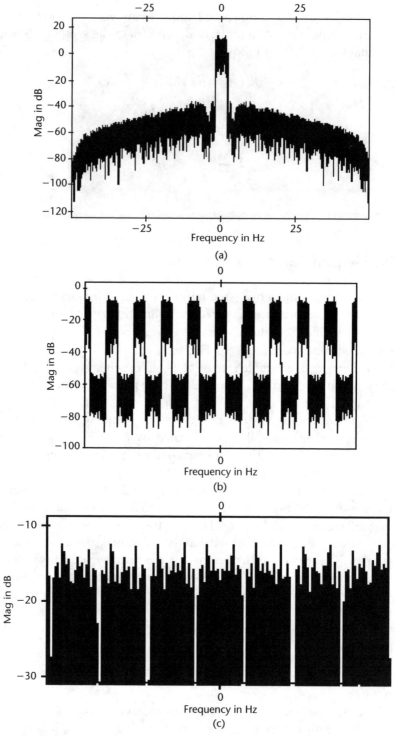

Figure 3.2 (a) Spectra of 2-Hz bandwidth signal; and (b) spectra of the same signal sampled at 10 Hz; and (c) $B = 2$ Hz bandwidth signal sampled at 4 Hz.

solution is to look at the spectra of the signal, which will usually decrease as the frequency increases, and pick some sampling rate where you can convince yourself

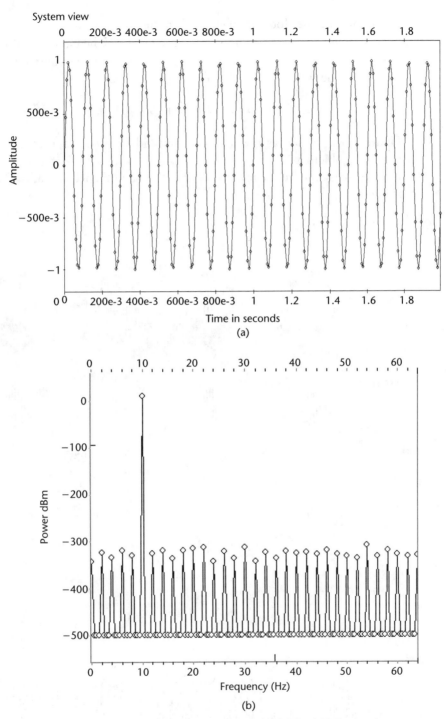

Figure 3.3 (a) 10-Hz sine wave sampled at 128 Hz; and (b) FFT of 10-Hz signal.

(and your boss!) that the small amount of aliasing does not harm your simulation. One such choice might be where the spectra at the alias frequency is, say, 48 dB down from the main center value. But other choices are possible.

Let us take a closer look at aliasing, with the signal being a cosine wave of some frequency f, which has been sampled at a rate $f_s = 1/T$. The bandwidth of the cosine wave is $2f$, from $-f$ to f. Now consider a 10-Hz cosine wave sampled at 128 samples/sec (Hz). The time domain samples can be written as

$$s(kT) = \cos(2\pi k 10/128) \quad T = 1/128$$

Figure 3.3(a, b) shows the time domain and frequency spectra of this signal. There is no problem at all.

Now increase the frequency to 63 Hz as shown in Figure 3.4 (a, b). There is still no problem.

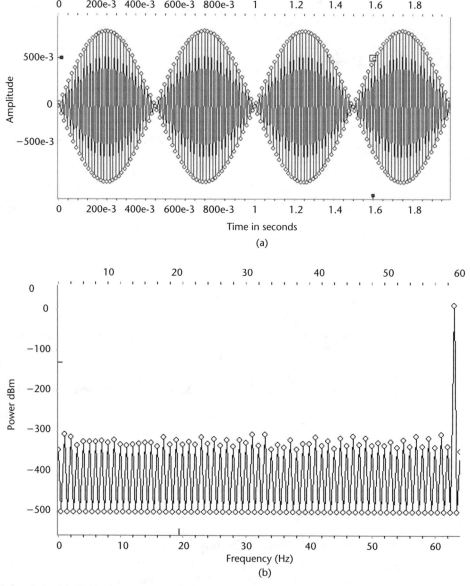

Figure 3.4 (a) 63-Hz sine wave sampled at 128 Hz; and (b) FFT magnitude of 63-Hz signal.

Now, try one more time with $f = 65$ Hz, as shown in Figure 3.5(a, b).

Compare the time plot and spectra here [Figure 3.5(a, b)] with the 63-Hz case [Figure 3.4(a, b)]. They are the same. What happened is that the 65-Hz signal "folded over" (aliased) around the $F_s/2$ value 64 Hz back to 63 Hz. Mathematically this can be written as follows:

$$s(k) = \cos[2\pi 65k/128]$$
$$= \cos[2\pi(128-63)k/128]$$
$$= \cos[2\pi k - 2\pi k 63/128]$$
$$= \cos[2\pi k 63/128]$$

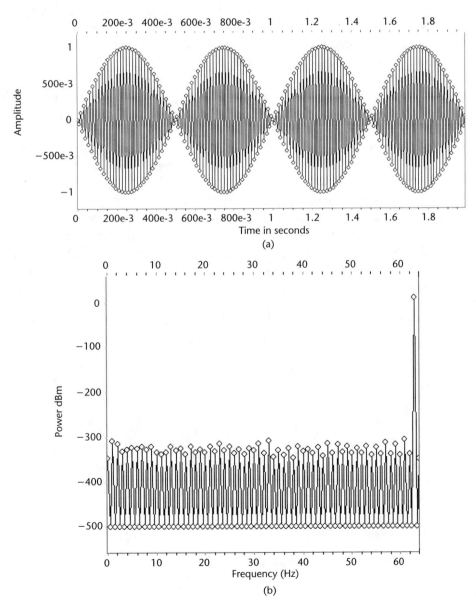

(a)

(b)

Figure 3.5   (a) 65-Hz sine wave sampled at 128 Hz; and (b) FFT magnitude of 65-Hz signal.

which is *identical* to the result for 63 Hz. So as we increase the frequency further, the spectral line will move back down to 0 Hz for an input of 128 Hz, and start moving back up, and so on. One point to emphasize is that once the aliasing has occurred, there is no undoing it.

3.3   Recovering the Signal from Its Samples

We now consider the inverse sampling operation. Given the sampled function $r_s(t)$, how can we reconstruct the original signal? From Figure 3.1(d), the issue is essentially how we connect the dots. This procedure is called *interpolation*, and there are many possibilities. The first and simplest way is simply to hold the previous sample until the next sample data point occurs, and hold that one until the next, and so forth. Figure 3.6 shows this simple operation. It is interesting to compare the original signal spectra shown in Figure 3.7 with that of the recovered signal spectra as shown in Figure 3.8. We see that the two agree at the low frequencies, but the interpolated spectra show a harmonic of the 10-Hz sample rate *Fs*. We could eliminate, or at least reduce this harmonic by following the hold operation with an appropriate low pass filter. A second idea is to linear interpolate; that is, connect the dots with straight lines. The result of this operation is shown in Figure 3.9, with the associated spectra shown in Figure 3.10. Note in Figure 3.10, the harmonic power is greatly reduced. We can continue this process by fitting parabolas over three point and so on. In textbooks on mathematics one can find a whole family of such functions known as Lagrange interpolators.

But the issue remains. Is there an ideal interpolation function that *exactly* recreates the signal? The answer is yes. The spectrum of the sampled function in Figure 3.2(b) shows the original spectra, Figure 3.2(a), repeated at multiples of the sample rate. Suppose that we multiply this spectrum $R_s(f)$ by the frequency domain function windowing:

$$H(f) = 1 \quad -f_s/2 \le f \le f_s/2$$
$$= 0 \quad \text{otherwise}$$

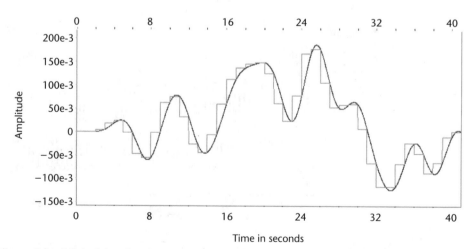

Figure 3.6   Original signal and recovered signal from *hold* interpolation.

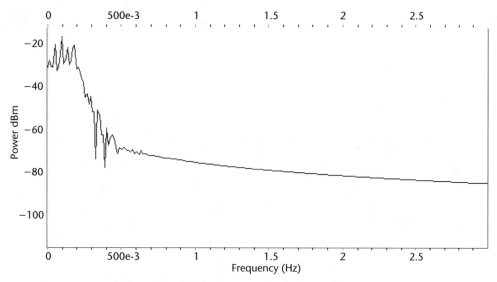

Figure 3.7   Original signal spectra.

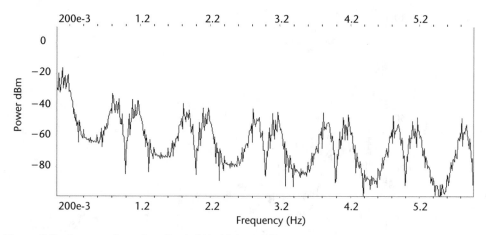

Figure 3.8   Recovered spectra after *hold last* interpolation.

The resulting spectrum, $R(f)$ is the original signal spectra and hence the Fourier transform of $R(f)$ is the original signal. The result is

$$s(t) = \int_{-\infty}^{\infty} R(f)e^{2\pi jft}\,dt = \int_{-\infty}^{\infty} R_s(f)H(f)e^{2\pi jft}\,df$$

$$= \int_{-\infty}^{\infty} r_s(\hat{t})h(t-\hat{t})\,dt$$

$$= \int_{-\infty}^{\infty} \sum_{k=-\infty}^{\infty} s(kT)\delta(\hat{t}-kT)h(t-\hat{t})\,d\hat{t}$$

$$= \sum_{k-\infty}^{\infty} s(kT)h(t-kT)$$

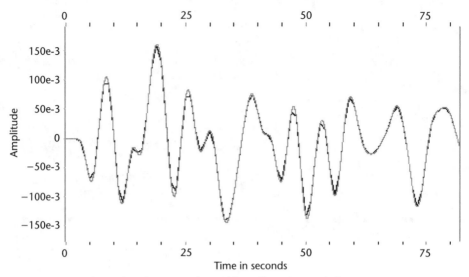

Figure 3.9   Original signal and recovered signal from linear interpolation

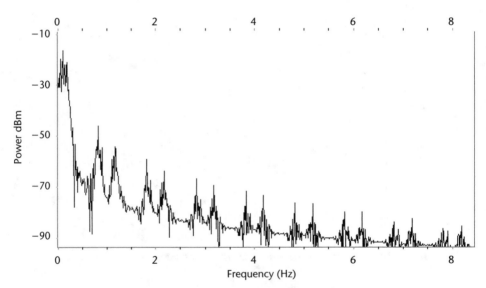

Figure 3.10   Power spectra of recovered signal using linear interpolation.

where $T = 1/f_s$ and

$$b(t) = \int_{-f_s/2}^{f_s/2} e^{2\pi jft} \, df$$
$$= F_s \, \text{sinc}(\pi t/T)$$
$$\text{sinc}(x) = \sin(x)/x$$

is the theoretically optimum interpolation function in that it is the *only* function that *exactly* recovers the original signal. Figure 3.11 shows this reconstruction for a portion of a complete signal. Notice that the sinc function has the property

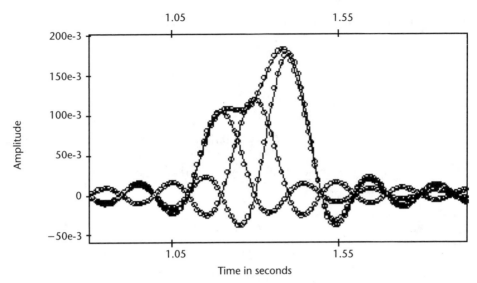

Figure 3.11   Reconstruction using sinc function interpolation.

$$\sin c^{+G}\left[B(t - kT)\right] = 0 \quad k \neq 0$$

Thus the sinc function does not interfere with the other data points at these times.

Now that we have learned that the sinc function is the theoretically optimum interpolation function, realize that *it does not exist in the real world*. This is due to the fact that it is derived from a perfectly band limited function, which implies as before: that the function exists for all time. In a simulation, this function must be truncated to some finite time mT. In fact, for small enough m, the truncated spectrum may perform worse than the other mentioned (suboptimum) interpolation functions that require far less processing.

3.4   Band-Pass Sampling

We just showed from the Nyquist theorem that the sample rate must be at least $B/2$. Now consider the following situation. We want to simulate an Army tactical radio. This radio has a bandwidth of 25 kHz and is transmitted on a 100-MHz carrier. In order to avoid aliasing the 100-MHz carrier, a sample rate of 200 MHz (–100 MHz to +100 MHz) would be required. This is a huge overkill since the information bandwidth of 25 KHz only requires a sample rate of 50 KHz, a factor of 3,200 less! One solution to this problem is to use baseband simulation techniques as described in Chapter 4. The other solution is the so-called band-pass sampling operation.

Recall that we also showed that the sampling operation is equivalent to mixing the signal with a series of oscillators whose frequencies are spaced by the sample rate f_s. In that discussion we started with a baseband signal with the resulting signal spectra being replicas of the baseband spectra translated up and down in frequency accordingly. The rules, however, do not change if the signal is a band-pass of some bandwidth B on a carrier f_0. Thus, if we sample this radio signal at f_s, the resulting translated sampled spectra f_k are at

$$f_k = 100MHz \pm kf_s$$

and

$$f_k = -100MHz \pm kf_s$$

Yet with all of these ups and downs in frequency, we only see that portion of the spectra in the region

$$-f_2/2 \leq f \leq f_s/2$$

By way of a specific example, Figure 3.12(a–c) shows the specifics for a $B = 2$-Hz signal on a 100-kHz carrier sampled at 36 MHz.

Figure 3.12(a) shows the pass band spectrum of the signal at the +/–100-kHz frequencies. The labels U0 and L0 refer to the corresponding spectral locations.

Figure 3.12(b) shows the resulting spectra after sampling the signal at 36 MHz. As stated, the upper signal at U0 = 100 kHZ is translated up to U-1 at

100 + 36 = 136 kHz, and is translated down to U-1 at 100 – 36 = 72 kHz,

U-2 to 100 – 72 = 36 kHz, and so on. A similar procedure applies to the original L0 at –100 kHz. The thing to observe is that none of these flying spectra crash into each other. This statement is not generally true with an arbitrary carrier frequency, signal bandwidth, and sample rate.

Figure 3.12(c) shows the final spectrum limited to +/–36 kHz. Not that the U-3 term has been translated down to –7.5 kHz while the L-3 term is at +7.5 kHz. In this case the spectrum is inverted

So far, we have ignored the signal bandwidth B. In the current example, if $B/2 <$ 5 MHz, then there would be no further issues with these flying spectra crashing into each other, or aliasing. It is clear that the best situation is when the sampled spectra centers up at $f_s/4$. This situation is commonly employed in actual digital hardware. First, this operation maximizes the bandwidth available to $f_s/2$. That is, the signal exactly fits into the sampled frequency space. Second, additional signal down conversion is greatly simplified. In Chapter 7 on demodulation we show that normally the first step is to I, Q down convert the signal from its bandpass location to baseband. The operation entails multiplying the signal with the sine and cosine of the location of the carrier frequency. But look what happens if the carrier is at $f_s/4$ and we then down convert with the same frequency. The resulting down conversion signals in this case are

$$I(k) = \cos\left[2\pi k(f_s/4)/f_s\right] = 0, \pm 1$$
$$Q(k) = \sin\left[2\pi k(f_s/4)/f_s\right] = 0, \pm 1$$

What a happy result! The down conversion process entails multiplying by [0, 1, –1]. There is no need for generating the trigonometric functions and employing time-consuming multiplications.

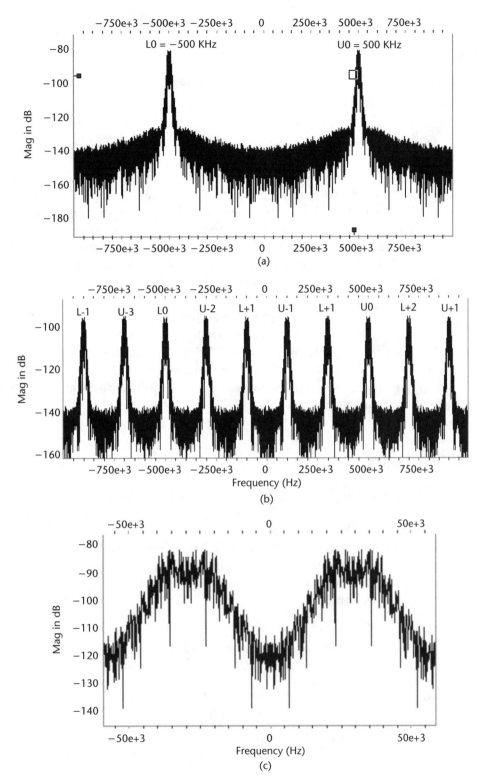

Figure 3.12   (a) Signal spectrum of a B = 2-Hz signal on a 100-kHz carrier; (b) the spectrum of the original signal sampled at 36 kHz; (c) final spectra of the band-pass sampled signal.

For a given carrier frequency f_0, what sample rates will result in an $f_s/4$ system? The basic relation is

$$f_0 - kf_s = f_s/4$$

or

$$f_s = 4f_0/[4k+1]$$

Table 3.1 evaluates this expression for the current example with $f_0 = 80$ MHz.

The particular value of k used depends, of course, on the bandwidth B of the signal to be processed. One word of caution needs to be mentioned here. The signal must be filtered to the bandwidth B *before* the sampling operation. While the signal itself is not affected by this operation, the system wideband noise floor is. Without this filter, the higher frequency noise components will fold back or alias, seriously affecting the simulation performance.

Finally, reconsider sampling the 80-MHz signal, only this time with a 30-MHz sample rate. Then the +80-MHz signal will show up at $80 - 3 \cdot 30 = -10$ MHz, while the signal at −80 MHz appears at +10 MHz. This phenomenon is called *spectrum inversion*. While no information is lost, some aspects of the demodulated signal such as phase or frequency may be inverted as well, requiring additional processing steps to recover the information.

3.5   Conclusion

In this chapter we laid the foundation for sampling the time function used to drive the simulation. We showed that the sampling operation is equivalent to mixing the signal with oscillators with frequencies at multiples of the sample rate. These operations lead us to the concept of aliasing and the all-important Nyquist sampling theorem. We also developed the optimum interpolation function for recovering the original signal from its sampled version. Finally, the useful concept of band-pass sampling was introduced.

Table 3.1   Sampling Values for an $f_s/4$ System at 80 MHz

k	f_s MHz	$f_s/4$ MHz	Max. bandwidth B MHz
0	320	80	160
1	64	16	32
2	35.55	8.89	17.78
3	24.62	6.15	12.31

Selected Bibliography

Pahlavan, K., and A. H. Leveseque, *Wireless Information Networks*, New York: John Wiley & Sons, 1995.

Proakis, J. B., *Digital Communications*, New York: McGraw-Hill, 1983.

Rappaport, T. S., *Wireless Communications, Principles and Practice*, Upper Saddle River, NJ: Prentice Hall, 2002.

Sklar, B., *Digital Communications, Fundamentals and Applications*, Upper Saddle River, NJ: Prentice Hall, 2001.

Steele, R., *Mobile Radio Communications*, London: Pentech Press Limited, 1992.

CHAPTER 4

Filters

In this chapter we take up how to implement filters in a simulation. We start out by developing the theories of the continuous Laplace or s transform, and its digital equivalent, the z transform. Then we develop the concepts for finite impulse response (FIR) filters and infinite impulse response (IIR) filters. Next we detail how one transforms from the s to z domain. Finally we take up some practical issues about running IIR filters in the z domain.

4.1 General Considerations

For a filter with a complex frequency domain transfer function $H(f)$, we can write it in the following useful form:

$$H(f) = A(f)e^{i\varphi(f)}$$

$A(f)$ is called the amplitude response, and $\varphi(f)$ is called the phase response. Both of these functions are real valued. In Chapter 2 we noted that for a real-time function, the FT must have the symmetry

$$H(-f) = H^*(f)$$

This in turn implies that $A(-f) = A(f)$, and $\varphi(-f) = -\varphi(f)$.

Now consider the response of such a filter to a pure cosine wave input frequency f_0. From basic FT, the output time response of the filter $y(t)$ is

$$y(t) = \text{Re}\left[\int_{-\infty}^{\infty} A(f)e^{i\varphi(f)}e^{2\pi ift}\delta(f - f_0)df\right]$$
$$= \text{Re}\left[A(f_0)e^{i\varphi(f_0)}e^{2\pi if_0 t}\right]$$
$$= A(f_0)\cos(2\pi f_0 t + \varphi(f_0))$$

The result emphasized the meaning of the gain, and phase (delay) of a filter. Now consider a somewhat more complicated case where the input $x(t)$ is two cosine waves with slightly different frequencies $f_0 \pm \Delta f$.

$$x(t) = \cos(2\pi f_0 t + 2\pi\Delta ft) + \cos(2\pi f_0 t - 2\pi\Delta ft)$$
$$= 2\cos(2\pi\Delta ft)\cos(2\pi f_0 t)$$

The result is a low frequency AM modulation of the carrier frequency f_0 by the cosine of frequency Δf. More generally it represents the common case of a narrow bandwidth signal modulated on a carrier. The modulating signal is called the envelope.

Now, the output response, $z(t)$, to this input is just the sum of the individual outputs of the form above:

$$z(t) = A(f_0 + \Delta f)\cos\left[2\pi(f_0 + \Delta f)t + \varphi(f_0 + \Delta f)\right]$$
$$+ A(f_0 - \Delta f)\cos\left[2\pi(f_0 - \Delta f)t + \varphi(f_0 - \Delta f)\right]$$

First we assume that $\Delta f \ll$ signal bandwidth, which allows us to write

$$A(f_0 \pm \Delta f) = A(f_0)$$

Furthermore, we can expand the phase terms as follows:

$$\varphi(f_0 \pm \Delta f) = \varphi(f_0) \pm \varphi'(f_0)\Delta f 2$$

After making these substitutions, we expand the cosine terms above and gather terms to yield

$$z(t) = 2A(f_0)\cos\left[2\pi\Delta f\left(t + \varphi'(f_0)/2\pi\right)\right]\cos\left[2\pi f_0 t + \varphi(f_0)\right]$$
$$= 2A(f_0)\cos\left[2\pi\Delta f\left(t - \tau_g\right)\right]\cos\left[2\pi f_0 t + \varphi(f_0)\right]$$

where we have the standard definition of the filter group delay

$$\tau_g = -\left[d\varphi/dg\right]_{f_0}\big/2\pi$$

Comparing this with the input $x(t)$, we see that the action of the filter is to shift the carrier phase term, while the envelope is shifted in time by the group delay.

Figures 4.1 and 4.2 shows this result with $\Delta f = 2$ Hz, and $f_0 = 100$ Hz. The filter is a three-pole Butterworth with 3-dB points at 95 and 105 Hz.

Figures 4.3 and 4.4 show these results for the Butterworth filter used to create Figures 4.1 and 4.2. The phase delay as shown in Figure 4.3 is 0 at 100 Hz. The group delay at 100 kHz shown in Figure 4.4 is about 0.056 sec. This is very close to

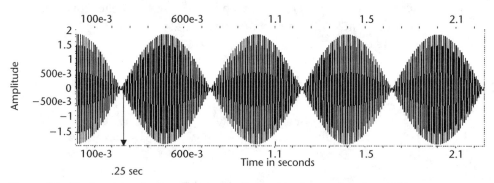

Figure 4.1   The input two-tone waveform. The first null in the envelope is at 0.25 sec.

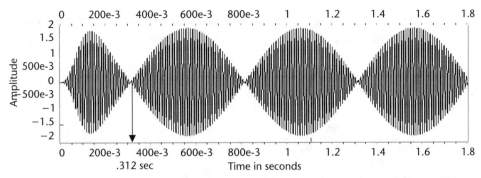

Figure 4.2 The filtered output two-tone signal. The first null in the envelope is now at 0.312 sec.

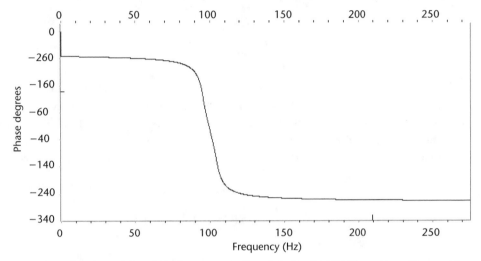

Figure 4.3 The phase delay of the Butterworth filter. It is 0 at $f = 100$ Hz and is antisymmetric with respect to that point.

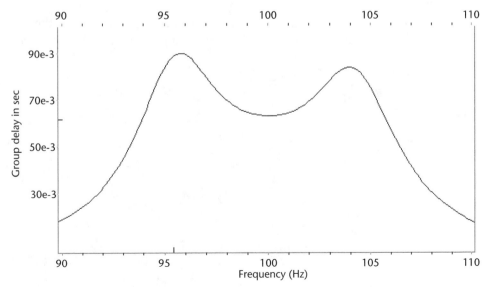

Figure 4.4 The group delay of the Butterworth filter. At the center 100 Hz it has the value 0.056 sec. Note the characteristic rabbit ears about the center.

the group delay as measured (to the best of our ability) in Figures 4.1 and 4.2 (0.312 − 0.25 = 0.062 sec), which can be seen by examination of the two signals.

4.2 The Laplace Transform

In Chapter 2 we developed the theory of the FT as an aid to implementing an LTI system. To proceed with our development, a second similar (but not identical) transform is required.

The Laplace transform (LT) is defined by the relations

$$H(s) = \int_0^\infty h(t)e^{-st}\,dt$$

$$h(s) = \int_{c-j\infty}^{c+j\infty} H(s)e^{st}\,ds$$

Note the similarity between the LT and the FT. There are two major differences. First, the lower limit on the time to s integral is 0, not −∞. The reverse transform integral is on a contour in the complex plane. This means that the LT is realizable in the real world. We can turn it on, so to speak, at any time we wish to call $t = 0$. It is instructive to emphasize this point by considering the LT of the derivative of a function:

$$H_d(s) = \int_0^\infty dh(t)/dt = sH(s) - h^+(0)$$

Where $h^+(0)$ is the initial condition of $h(t)$ taken at zero from the + side. Appendix A gives a short table of Laplace transforms.

What this tells us is that the LT can incorporate initial conditions of a system where the FT cannot. More importantly, the impulse response of a signal derived from an LT includes the initial transient of the system as well as the steady state response. To illustrate this in detail, consider a differential equation of the form

$$dy(t)/dt + y(t) = \sin(\omega t)$$

Taking the LT of both sides gives

$$LT\big[dy(t)/dt + y(t)\big] = LT\big(\sin(\omega t)\big)$$

$$LT\big[dy(t)/dt\big] + LT\big[y(t)\big] = LT\big(\sin(\omega t)\big)$$

$$sH(s) - h^+(0) + H(s) = \omega/\big(s^2 + \omega^2\big)$$

$$(s+1)H(s) = h^+(0) + \omega/\big(s^2 + \omega^2\big)$$

$$H(s) = \big[h^+(0) + \omega/\big(s^2 + \omega^2\big)\big]/(s+1)$$

$$= \big[\omega/\big(1 + \omega^2\big)\big]\big[1/(s+1) - (s-1)/\big(s^2 + \omega^2\big)\big]$$

The time domain result $y(t)$ is now

$$y(t) = \left[\omega e^{-t} + \sin(\omega t) - \omega\cos(\omega t)\right]/\left(1+\omega^2\right)$$

where we have taken $h^*(0) = 0$. It is easy to verify that $y(o) = 0$ as required.

The output $y(t)$ consists of two parts: an exponentially dying transient and a steady state sinusoidal term (which is an amplitude and phase shifted version of the input sine wave). This result is shown in Figure 4.5.

Note that as time goes to infinity, the steady state of $y(t)$ is the exact result that would have been obtained by using the FT.

A second difference is that the FT transform variable, f, is a real number, with the complex number j explicitly shown. In the LT, s is considered to be a complex number of the form

$$s = \sigma + j\omega = \sigma + j2\pi f$$

From this we can identify the imaginary axis of the s domain transform with the frequency domain parameter of the FT.

The inverse LT is

$$h(t) = \int_C H(s)e^{st}\, ds$$

where the integration is along the imaginary axis in the s plane. Tables of LTs can be easily found that contain nearly every useful pair that comes into normal analysis.

4.3   Poles, Zeros, and Stability

The LT of a simple exponential $s(t) = e^{at}$ (with no restriction on a) is

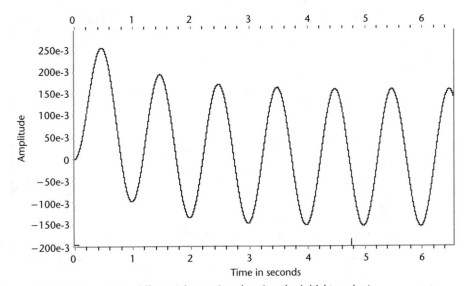

Figure 4.5   Solution to the differential equation showing the initial transient.

$$H(s) = 1/(s-a)$$

A pole in the transform function is where the denominator goes to zero, or equivalently where $H(s)$ goes to infinity. In this case the pole is at $s = a$ in the Laplace domain plane. If a is, in fact, positive, the pole is in the right half of the plane $\sigma > 0$. The exponential in time is then increasing without limit, which is known as an unstable signal. If a is negative, the pole is in the left half plane. The exponent is decaying in time, which is a stable condition. The precise case $\sigma = 0$ is called neutral stability. From this discussion we arrive at the standard definition of a stable system as one where all of the poles are in the left half plane, including the axis.

The opposite of a pole is a zero. This is a value of s such that $H(s) = 0$. In the function

$$H(s) = s + 2$$

the pole is at $s = -2 + 0j$ in the Laplace plane.

In the general linear system, the Laplace function is of the form

$$H(s) = \frac{N(s)}{D(s)} = \frac{\prod_{k=1}^{n}(s - z_k)}{\prod_{k=1}^{m}(s - p_k)}$$

That is a ratio of two polynomials. The superscripts n and m in the formula represent the number of poles and zeros describing the filter. For example, the filter

$$H(s) = (s+2)/(s^2 + 2s + 2)$$

has one zero at $s = -2$, and two poles at $s = -1 \pm j$ (for a real-time function, the poles must occur in complex conjugate pairs). Figure 4.6 shows the standard graphic used to describe the filter 0 = zero; x = pole.

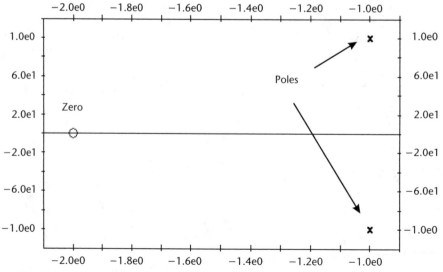

Figure 4.6   Pole, zero representation of a filter in the s plane.

4.4 The Two Worlds of a Filter

4.4.1 Infinite Impulse Response filters

Consider the simple one pole LT pair

$$H(s) = 1/(s+a) \leftrightarrow y(t) = e^{-at}$$

The time exponential is the impulse response of the filter. Note that $y(t)$ *never* reaches zero. Admittedly, it becomes so small that it may be considered zero for all practical purpose, but in the strict mathematical sense it never gets there. What we see in general is that any filter that has poles will also have a time impulse response that never goes to zero. This is the basis of an IIR filter. Most of the common filters such as Butterworth, Bessel, Elliptic, Chebychev, and the like are IIR filters. They can be implemented on the bench (analog world) via suitable lumped constant inductor resistor capacitor (LRC) filter as shown in Figure 4.7.

This is a simple circuit with the transfer function

$$H(s) = V_{out}(s)/V_{in}(s) = 1/(LCs^2 + RCs + 1)$$

where the output voltage is taken across the capacitor.

If we desire a two pole Butterworth filter, we require: LC = 1 and RC = 1.414. One solution is to take $L = 1/1.414$ h, R = 1 ohm, and C = 1.414 farad, and then the above reduces to

$$H(s) = V_{out}(s)/V_{in}(s) = 1/(s^2 + 1.414s + 1)$$

This is the transfer function of a two pole Butterworth filter with a 1 ra/sec 3-dB point. Other standard filters have similar, but more complicated circuits.

The most common IIR filters are:

- Butterworth;
- Bessel;
- Chebychev (I and II);
- Elliptic;
- Gaussian.

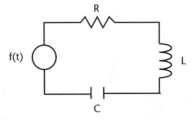

Figure 4.7   Simple LRC circuit that implements a two pole Butterworth filter.

4.4.2 Finite Impulse Response Filters

An FIR filter has an impulse response that is confined to some specific time extent. The common structure is shown in Figure 4.8.

The time transfer function of this filter is simply

$$y(t) = \sum_{k=1}^{4} c_k x(t - k\tau)$$

and the LT of the filter is simply

$$H(s) = \sum_{k=1}^{4} c_k e^{-ks\tau}$$

If we send an impulse into this filter, the output will be the same impulse, weighted by the coefficients c_k, appearing at multiples of the delay τ. Once the impulse passes the last delay stage there is no more output, hence the title FIR filter.

Surface acoustic wave (SAW) devices are a classic example of an FIR filter operating in the analog or continuous domain. These filters use overlapping electrodes on a piezoelectric substrate to implement the desired band-pass response.

4.5 The z Transform

Now that we have gone through the development of the Laplace and Fourier transforms as they apply to the representation of filters, the reader is again reminded that such representations do not exist on a computer. What we need is the digital equivalent of the FT. This is called the z transform.

The z transform may be considered as the sampled or digital equivalent of the continuous LT. We start with a continuous signal $x(t)$ and sample it at a rate $1/T$. A typical result was shown in Figure 3.1(d). The mathematical expression is

$$x_s(t) = \sum_k x(kT)\delta(t - kT)$$

Now take the LT of the above to get

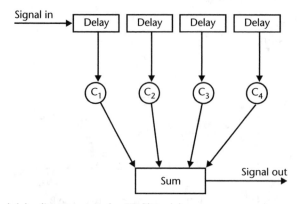

Figure 4.8   Tapped delay line structure for FIR filter delay $= \tau$.

$$LT\left[\delta(t-\tau)\right] = e^{-s\tau}$$

$$X(s) = \sum_k x_k e^{-ksT}$$

If we denote $z = e^{sT}$, then the z transform $X(z)$ is defined as

$$X(z) = \sum_k x_k z^{-k}$$

As an example consider the exponential e^{-at}. Then we have

$$X(z) = \sum_{k=0}^{\infty} e^{-kaTz^{-k}} = \sum_{k=0}^{\infty}\left(e^{-aT}z^{-1}\right)^k$$

$$= 1/\left(1 - e^{-aT}z^{-1}\right)$$

Now define a z domain filter with impulse response $H(z)$, which has the general form

$$H(z) = \sum_{k=0}^{\infty} c_k z^{-k}$$

What is the output of this filter $Y(z)$ for an input function $X(z)$? In Chapter 2 we derived the time domain result for the input output of a continuous filter by actually starting with a discrete sampled signal and then letting the sample rate go to infinity. The summation then went over to an integral. But here we do not need to go that far. Following the procedure described in Chapter 2, we immediately have

$$Y(z) = H(z) * X(z)$$

4.6   From s to z

The Laplace variable s, as mentioned, is complex. Thus, the z variable is complex as well. What we have is a mapping from the s to z complex planes. A little analysis shows that the imaginary $j\sigma$ axis in the s domain maps to the unit circle in the z domain. Furthermore the left hand side ($\sigma < 0$) of the Laplace domain is mapped into the unit circle in the z domain.

The problem that we face when simulating any transfer function (filter) that is defined in the s domain is how to convert the design to the z domain that is the basis for the implementation algorithm? In other words, we need to define some function of z and let $s = f(z)$. By definition $z = e^{sT}$, then a logical choice would be $s = f_s \ln(z)$. But we cannot implement $\ln(z)$ in any reasonable manner so another $f(z)$ must be found. There is no unique solution here. All we can do is set up some criteria for $f(z)$ which takes an $H(s)$ into an $H(z)$ which is representative of the original $H(s)$.

There is no guarantee that an arbitrary $f(z)$ will work. One mandatory requirement is that $f(z)$ maps the entire left hand side of the s plane into the unit circle of the z plane. Without this, the transformed filter may be unstable. Consider an example using a simple derivative

$$LT\big[dy(t)/dt\big] = sY(s)$$

Now the derivative $dy(t)/dt$ can be numerically approximated by

$$dy(t)/dt \cong \big[y(t_k) - y(t_{k-1})\big]/dt$$

Now apply the z transform to the RHS of the above to obtain

$$ZT\big[y_k - y_{k-1}\big]/dt = Y(z)\big[1 - z^{-1}\big]f_s$$

Combining results, we obtain the desired transform

$$x \leftrightarrow f_s\big[1 - z^{-1}\big]$$

It is easy to show that this relation maps the left half plane (LHP) of s to a circle of radius 1/2, centered at z = 1/2, as shown in Figure 4.9.

Thus the stability criterion is satisfied. However, further investigation shows that the resulting z domain filter does not preserve the properties of the original $H(s)$, especially when $H(s)$ starts to have significant frequency components that approach the sample rate fs.

The $f(z)$ that is usually used is the bilinear transform defined by

$$s = 2f_s\left(1 - z^{-1}\right)\big/\left(1 + z^{-1}\right)$$

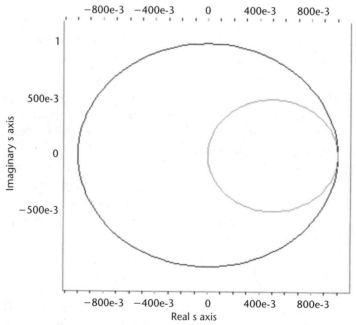

Figure 4.9 The z plane mapping of the simple derivative s to z transformation. The mapped circle is within the unit z domain circle.

This transform maps $s = j\omega$ to the unit circle in the z plane. Furthermore, the resulting z domain filter retains most (but not all) of the properties of the original filter.

For example,

$$H(s) = 2\pi/(s + 2\pi)$$

$$s = 2f_s\left(1 - z^{-1}\right)/\left(1 + z^{-1}\right)$$

$$H(z) = 2\pi/\left[2f_s\left(1 - z^{-1}\right)/\left(1 + z^{-1}\right) + 2\pi\right]$$

$$H(z) = 2\pi\alpha\left(1 + z^{-1}\right)/\left(1 + \beta z^{-1}\right)$$

$$\alpha = 1/(2\pi + 2f_s)$$

$$\beta = \alpha(2\pi - 2f_s)$$

There are other ways to obtain this transform other than just pulling the proverbial rabbit out of the hat, as we have done. One way is to return to the original idea $s = f_s\ln(z)$. Only we now observe the following expansion of $\ln(z)$:

$$\ln(z) = 2\sum_{k=1}^{\infty}\frac{\left(1 - z^{-1}\right)}{(2k - 1)\left(1 + z^{-1}\right)}$$

If we keep the first term in the expansion, we recover the basic result. This result immediately begs the question: What if I keep more than one term in this expansion, would things improve in some sense? (The author is not sure, but thinks this would be a good thesis project.) It would be a more complicated transformation, so the z domain order of the filter would increase. This, of course, increases the required simulation speed or increases the complexity of direct implementation.

A second idea is based on the numerical integration algorithm, known as Simpson's rule, illustrated in Figure 4.10.

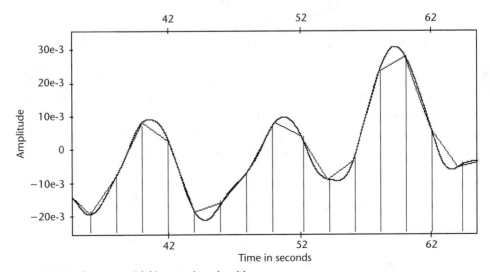

Figure 4.10   The trapezoidal integration algorithm.

First start with the basic relation

$$LT\left[\int y(t)dt\right] = Y(s)/s$$

From Figure 4.10 we have the trapezoidal integration algorithm

$$\int y(t)dt = 5_s\left[(y_0 + y_1) + (y_1 + y_2) + (y_1 + y_2)..\right]/f_s$$

Now take the z transform of the RHS of the above as we did before to obtain

$$\int y(t)dt = 5_s\left[(y_0 + y_1) + (y_1 + y_2) + (y_1 + y_2)..\right]/f_s$$

$$H(z) = 5\left[(1 + z^{-1}) + z^{-1}(1 + z^{-1}) + z^{-2}(1 + z^{-1})+..\right]Y(z)/f_s$$

$$= 5(1 + z^{-1})\left[1 + z^{-1} + z^{-2} + z^{-3}+..\right]Y(z)/f_s$$

$$= \left[(1 + z^{-1})/2f_s(1 - z^{-1})\right]Y(z)$$

From this we can write the transformation

$$s = 2f_s(1 - z^{-1})/(1 + z^{-1})$$

which is the basic transform again.

4.7 Frequency Warping

In the s domain, the filter described in the last section

$$H(s) = V_{out}(s)/V_{in}(s) = 1/(s^2 + 1.414s + 1)$$

has a 3-db half power point at $f = 1$ Hz. Figure 4.11 shows the impulse gain response of this filter via the detailed $H(z)$ with $f_s = 5$ Hz.

A close examination of this filter shows something strange. The gain at $f = 1$ Hz is not −3 dB but −3.64 dB. The actual 3-dB point is at $f = 0.891$ Hz. This phenomenon is called warping. To see why this happens, return to the basic bilinear transformation and substitute $z = e^{j\Omega}$ and $s = j\omega$. This substitution maps the frequency f = ù/2ð in the s domain to the frequency $f_z = \Omega f_s/2\pi$ in the z domain. The result is

$$2\pi f = 2f_s \tan(\Omega/2)$$
$$= 2f_s \tan(\pi f_z/f_s)$$

where f_z is the corresponding frequency resulting from the transformation. Now as the sample rate becomes large, we have $f = f_z$. Substituting $f = 1$ into the above results in the $f_z = 0.891$ Hz frequency already found.

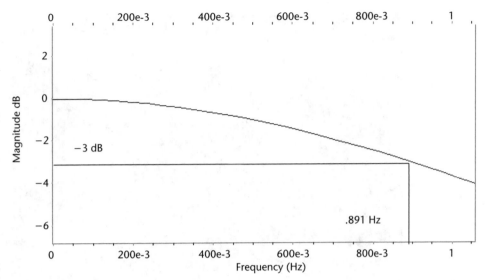

Figure 4.11   Gain response of the simple one pole filter. The 3-dB point is *not* at the 1-Hz design point.

For an arbitrary $H(s)$ there is not much that can be done to correct this warping. But in the case of s domain filters, we control both ends of the problem. The trick is to calculate via the above equation what frequency in the s domain must be used in order to arrive at the desired frequency as executed in the z domain. This procedure is called prewarping. In the case at hand, if we prewarp to 1.156 Hz, the executed 3-dB point will be the desired 1 Hz

Many design procedures for FIR filters exist in the literature. The general idea is to provide the information on the filter as shown in Figure 4.12. Here the "frequencies" are entered as fractions of the sample rate. The transfer function information is the pass band gain, the transition band, and the ultimate rejection. In Figure 4.12 the design goal is flat 0 dB attenuation to the start of the transition frequency at 0.1 relative to the same rate. The end of the transition region is at relative frequency of 0.2, and from there the filter out of band rejection is 40 dB. Most algorithms use some sort of mean square error to arrive at a solution as close to the desired solution as possible. The goal is to compute the best fit to the desired with the minimum number of taps. In Figure 4.12, it is estimated that the filter can be realized with 21 taps. Now if we make things more stringent then, the number of taps goes up. For example, changing the first transition frequency to 0.1 to 0.11 Hz would increase the number of taps to 271. This is due to the fact that the transition bandwidth went from 0.1 to 0.09 relative. The sharper the transition, the greater the number of taps required.

4.8   The IIR Filter

In contrast to the FIR filter, the impulse response of an IIR filter is not confined to a specific time extent. For a stable system, the amplitude decreases with time, and at some value it may be regarded as negligible. We have already encountered such a filter $H(s) = 1/(s + 1)$, which has an impulse response e^{-t}.

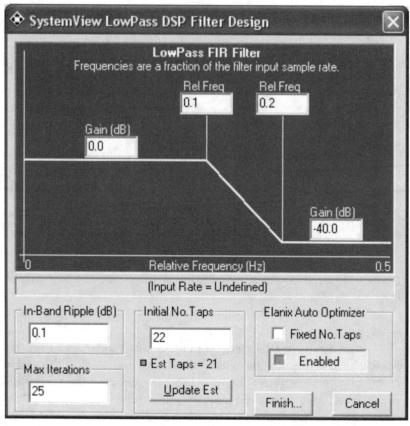

Figure 4.12 Typical design form for a FIR filter.

The design steps to implement an IIR filter are as follows:

1. Select the type of filter and the order desired.
2. From tables obtain the s domain poles listed for a 1-ra/sec cut-off frequency.
3. Prewarp this filter as required.
4. In practice there exist tables that specify these poles for a LPF filter with 3-dB point of 1 ra/sec (not 1 Hz). From there, one applies the following transformations to achieve the other filter types:

   - Low-pass low-pass: $s \rightarrow s/\omega_u$
   - Low-pass high-pass: $s \rightarrow \omega_u/s$
   - Low-pass band-pass: $s \rightarrow \left(s^2 + \omega_1\omega_u\right)/\left[s(\omega_u - \omega_1)\right]$
   - Low-pass bandstop: $s \rightarrow s(\omega_u - \omega_1)/\left(s^2 - \omega_u\omega_u\right)$

   where ω_u and ω_1 are the upper and lower cut-off frequencies (ra/sec), respectively.

5. Apply the bilinear transform. Clear fractions to obtain the requisite ratio of two z domain polynomial $N(z)/D(z)$.

Example

We consider a simple design that illustrates these steps while minimizing the algebra required (which is still messy).

Design a first-order Butterworth band-pass filter with low frequency of $\omega = 10$ Hz, high frequency of $\omega_u = 12$ Hz, and with a system sample rate of $f_s = 100$ Hz.

The transfer function with respect to a 1-ra/sec low-pass cut-off is

$$H_{LP}(s) = 1/(s+1)$$

The first step is to apply the low-pass to band-pass transformation to this function

$$s \rightarrow \left(s^2 + \omega_l \omega_u\right)/S(\omega_u - \omega)$$

The result is

$$H_{BP}(s) = a_0\, s/\left(s^2 + a_1 s + a_2\right)$$

where

$$a_0 = \omega_u - \omega$$
$$a_1 = a_0$$
$$a_2 = \omega_u \omega_l$$

A simple consistency check is to apply dimensional analysis to the results. Notice the three terms of the denominator. The *s* term has dimensions (ra/sec); the s^2 term has the same dimensions since a_1 has dimensions of ra/sec, and the constant term agrees since a_2 also has dimensions of (ra/sec). This type of analysis goes a long way toward catching algebraic errors!

Next we apply the bilinear transformation to get

$$H_{BP}(z) = 2a_0 f_s \left(1 + z^{-1} - z^{-2} - z^{-3}\right)\Big/\left(b_0 + b_1 z^{-1} + b_2 z^{-2}\right)$$

where the coefficients are

$$b_0 = 4f_s^2 + 2f_s a_1 + a_2$$
$$b_1 = -8f_s^2 + 2a_2$$
$$b_2 = 4f_s^2 - 2f_s a_1 + a_2$$

4.9   General Implementation of a Filter of Form *N(z)/D(z)*

Finally, after all of the required manipulations, the final result is an input output relation

$$Y(z) = N(z)X(z)/D(z)$$

The question is, how do we implement this relation to obtain a final time domain signal? We proceed by writing the above in the form.

$$D(z)Y(z) = N(x)X(z)$$

$$\sum_{k=0}^{m} b_k z^{-k} Y(z) = \sum_{k=0}^{n} a_k z^{-k} X(z)$$

$$b_0 = 1$$

By the rules of the z transform, we have the following equivalence:

$$z^{-p} Y(z) \leftrightarrow y(pT) = y_p$$

So the final time domain representation is

$$y_1 = -\sum_{k=1}^{m} b_k y_{l-k} + \sum_{k=1}^{n} a_k x_{l-k+1}$$

4.10 Practical Consideration in IIR Filter Simulation

Consider a 2N pole IIR filter with a transfer function of the form

$$H(s) = 1 \Big/ \left[\left(s^2 + a_1 s + b_1 \right)^* .. ^* \left(s^2 + a_N s + b_N \right) \right]$$

Now if this filter is stable in the s domain, then the resulting z domain filter

$$H(z) = N(z)/D(z)$$

is stable as well. Or is it? In all of our discussions regarding the finite nature of parameters in a computer, we have not considered the ramifications that a numeric coefficient is quantized as well. On a PC, it is possible to quantize a floating point to double precision, which is characterized by double precision floating point that represents the signal fraction into 52 bits and the overall amplitude into 12 bits = 11 numeric plus 1 bit sign. Now this is quite formidable, but in the case of the IIR it may not be enough. If the order N of the filter is large and the frequency cut-off is much less (1%) of the system sample rate, this numeric quantization can make the system unstable. What happens is that the zeros of $D(z)$, which are all inside the unit circle with infinite precision, can slip across the circle with the finite numeric math, making the filter unstable. This phenomenon is aggravated if the algorithm is executed on a digital signal processing (DSP) chip where memory is at a premium and severely limits the number of bits used to represent the coefficient.

One solution to this problem is to break the large filter into smaller pieces and execute each piece separately. Figure 4.13 shows the basic concept for a four pole filter. The two sections can be combined into one complete filter, or each section can be executed separately. The differences between the two are seen in the following equations:

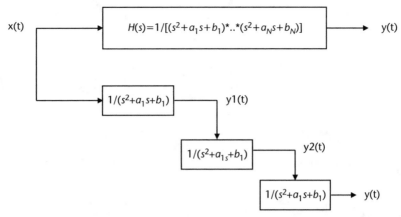

Figure 4.13 Alternative implementation of a filter. The section-by-section version has fewer stability problems due to finite representation of the filter coefficients.

$$Y(s) = H_1(s)H_2(s)X(s)$$
$$Y_1(s) = H_1(s)X(s)$$
$$Y_2(s) = H_2(s)Y_1(s)$$

In a perfect world, both implementations shown are mathematically equivalent. But the piecewise solution shown greatly mitigates the effects of the finite precision arithmetic.

4.11 Conclusion

In this chapter we developed the theory of filters. We started with the general concept of amplitude gain, phase delay, and group delay. The Laplace transform was introduced. Unlike the Fourier transform, the LT can account for initial system transients as well as the steady state obtained from the FT. The stability of filters with respect to the s plane was presented. The world of filters was split into IIR and FIR filters. The IIR filter is usually the result of an appropriate LRC circuit transfer function. As stated many times in this book, "continuous" does not exist in a computer. Sampling must be performed. To this end, we introduced the discrete version of the LT, known as the z transform. The bilinear transform was developed as the connector between the s and z domains. Finally we showed that the finite representation of the filter coefficients could make an otherwise stable filter, unstable. It was noted that by breaking up a large order filter into a series of shorter order ones, this stability issue could be significantly reduced.

Selected Bibliography

Rabiner, R. L., and B. Gold, *Theory and Applications of Digital Signal Processing*, Englewood Cliffs, NJ: Prentice-Hall, 1973.

Zverey, A. I., *Handbook of FILTER SYNTHESIS*, New York: John Wiley & Sons, 1967.

CHAPTER 5

Digital Detection

In this chapter, we develop the basic concepts used in digital detection theory. We start with a concept called a vector channel and arrive at the concept of vector distances and decision boundaries. Following this, we extend the analysis to that of time domain signals. Several optimum receiver architectures are derived and outlined.

5.1 The Vector Channel

We start with the simplest example, shown in Figure 5.1. A logical [1] is encoded as a value A1, and a logical [0] as a value A2. Between the source and receiver, the channel is AWGN, which is modeled as a Gaussian random variable (GRV).

The detection rule is to choose between A_1 and A_2, according to the conditional probability that

$$P[A_1|r] \geq P[A_2|r] \quad \text{choose } A_1$$
$$P[A_2|r] < P[A_1|r] \quad \text{choose } A_2$$

The standard trick is to employ Bayes theorem to invert the arguments of the conditional probability

$$P[r\pi A_k] = P[r|A_k]P[A_k]$$
$$P[A_k\pi r] = P[A_k|r]P[r]$$
$$P[r\pi A_k] = P[A_k e^{tr}]$$

We can now write

$$P[A_k|r] = P[r|A_k]P[A_k]/P[r]$$

Finally, we find a boundary value $\underline{r}$ according to the rule

$$P[A_1|r] = P[A_2|r]$$
$$P[r|A_1]P[A_1]/P[r] = P[r|A_2]P[A_2]/P[r]$$
$$P[r|A_1] = P[r|A_2]$$

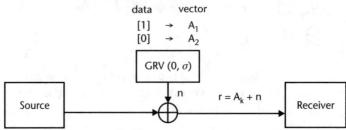

The channel adds noise. What is the structure of the receiver for optimum recovery of the information?

Figure 5.1 Basic binary vector channel detection diagram.

We have made the logical assumption that each message has equal probability of occurrence.

What have we gained by all of this manipulation? A lot, because we can write the expression for $P[r|A_k]$. The received value r is the sum of the message and the GRV noise term. It then follows that r is a GRV as well. So the only thing we need to know is the mean and variance of r, specifically

$$E(r) = E(A_k + n)$$
$$= E(A_k) + E(n)$$
$$= A_k$$
$$E\left[(r - A_k)^2\right] = E\left[n^2\right] = \sigma^2$$

Since the noise is a GRV, we can write

$$P(r|A_k) = \frac{1}{\sqrt{2\pi\sigma^2}} e^{-(r-A_k)^2/2\sigma^2}$$

Combining this equation with the criteria for the decision boundary reduces to

$$\frac{1}{\sqrt{2\pi\sigma^2}} \exp\left[-(\hat{r} - A_1)^2/2\sigma^2\right] = \frac{1}{\sqrt{2\pi\sigma^2}} \exp\left[-(\hat{r} - A_2)^2/2\sigma^2\right]$$

which further simplifies to the simple result

$$(\hat{r} - A_1)^2/2\sigma^2 = (\hat{r} - A_1)^2/2\sigma^2$$
$$\hat{r} = (A_1 + A_2)/2$$

Figure 5.2 shows the situation described. The decision boundary r is where the two individual PDF functions intersect. The peak of the two PDFs are at A_1 and A_2, respectively. So if $r > \hat{r}$ we choose A_2, and if $r < \hat{r}$ we choose A_1.

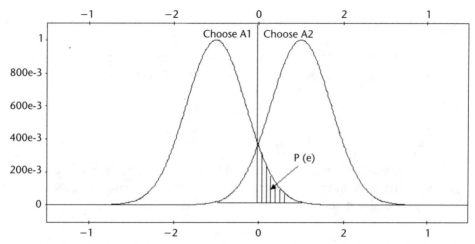

Figure 5.2 Decision boundaries for two amplitude vector channels. The boundary is where the two PSDs are equal.

Now the big question: What is the probability of error occurring in this decision process? An error occurs when, say, A_1 is transmitted and the added noise is sufficient to make $r > r$. The probability of such an error is simply

$$P(e) = \int_{\hat{r}}^{\infty} \exp\left[-(x - A_1)^2 / 2\sigma^2\right] dx / \sqrt{2\pi\sigma^2}$$

The area of integration is shown as the shaded area in Figure 5.2. A common term used here is the "tail of the Gaussian." This integral can be evaluated as a special function that is related to the complementary error function erfc(x) as

$$P(e) = Q(\beta) = \int_{\beta}^{\infty} \exp\left(-x^2 / 2\right) dx / \sqrt{2\pi} = \frac{1}{2} erfc\left(\beta / \sqrt{2}\right)$$

$$\beta = (A_2 - A_1) / \sigma$$

Let us slow down here, for this result is *extremely* important. Notice that $P(e)$ depends only on the *difference* between A_1 and A_2, not their individual values. In short the combination [$A_1 = 1$, $A_2 = 2$], gives the same error performance as [$A_1 = 1,000$, $A_2 = 1,001$]. This difference is seen as the *distance* between the two. This concept of distance is fundamental to all digital communication theory.

We continue with a more complicated example. We now wish to send one of four equally likely messages. Each message would then correspond to a 2-bit data pattern. Each message is represented by a two-dimensional vector (in the previous result the scalars A are just one-dimensional vectors). Specifically we have

$$m_1 \rightarrow [x_1, y_1] = [1,1]$$
$$m_2 \rightarrow [x_2, y_2] = [-1,1]$$
$$m_3 \rightarrow [x_3, y_3] = [-1,-1]$$
$$m_4 \rightarrow [x_4, y_4] = [1, -1]$$

This representation corresponds to a QPSK modulation format. Figure 5.3 shows this representation.

In the transmission process each component of the message is corrupted by an independent sample of AWGN, so the received vector is

$$r_x = x_k + n_x$$
$$r_y = x_y + n_y$$

Proceeding as we did before and by simple symmetry, we find that the decision boundaries are simply the coordinate axis. Again by symmetry, the smallest (nearest neighbor) distance

$$d_k^2 = \left(r_x - x_k\right)^2 + \left(r_y - y_k\right)^2 \quad k = 1,2,3,4$$

from the received vector $[r_x, r_y]$ to a possible code word is the same as just choosing the code word by the quadrant that the received signal falls in. This decision process is simplified since we only need to know the algebraic signs of r_x and r_y.

The above example was chosen since it represents a common decision problem relating to QPSK, and to the simplicity of the mathematics required to establish the decision boundaries. However, the general case would be treated in exactly the same manner only with some more tedious algebra.

5.2   The Waveform Channel

Of course, we do not send vectors over the air; we send some waveform $s(t)$. We send one of M time functions

$$s_k(t) \quad 0 \leq t \leq T$$
$$1 \leq k \leq M$$

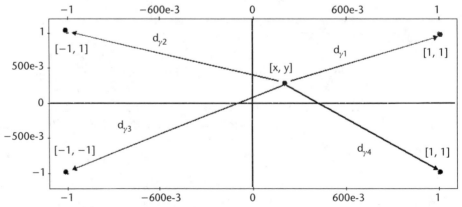

Figure 5.3   Optimum decision boundaries for the 4 vectors shown. This example is valid for QPSK modulation.

   The received waveform is sampled at a rate $m\Delta t = T$. Each sample $r(l) = s_k(l) + n(l)$ is the original signal plus an additive noise term. Each such noise term is independent from one sample to the next. Following the same steps as before, the PDF of the received signal is an m-dimensional Gaussian distribution. The boundary decisions are obtained again by equating the exponent of this distribution. The result is for two signals k and p

$$\sum_{l=1}^{m}\left[r(l) - s_k(l)\right]^2 = \sum_{l=1}^{m}\left[r(l) - s_p(l)\right]^2$$

   If we multiple the above by Δt, and take the limit as $\Delta t \to 0$, the sum goes over to an integral yielding

$$\int_0^T\left[r(t) - s_k(t)\right]^2 = \int_0^T\left[r(t) - s_p(t)\right]^2$$

   Now we make the common assumption that the energy of each signal is the same.

$$\int_0^T s_k^2(t)\,dt = \text{Energy, all } k$$

yields the desired result:

$$\int_0^T r(t)s_k(t)\,dt = \int_0^T r(t)s_p(t)\,dt$$

   This is a basic result. What is says is the optimum receiver correlates the received signal $r(t)$ with all possible basis wave functions $s(t)$, and the one that produces the largest value is chosen as the transmitted message. This is called a correlation receiver, and it is shown in Figure 5.4.

5.3   The Matched Filter Receiver

A basic problem in detection is illustrated in Figure 5.5. A known signal $s(t)$ is corrupted by noise. The question is: What is the optimum filter, $H(f)$, that maximizes the SNR at the output at some time T?

   There is a formal derivation of the answer, but let us try and sneak up up on it by using some simple logic. Let $S(f)$ be the PSD of $s(t)$. Suppose $S(f)$ has no content $[S(f) = 0]$ from, say, 10 to 20 Hz. Then there is no reason for $H(f)$ to pass any frequency in this range since it would only contain noise. Conversely, if $S(f)$ is large (10 to 20 Hz), $H(f)$ should be large as well. The logical conclusion is that $H(f)$ should look as much like $S(f)$ as possible, which leads to the actual answer:

$$H(f) = S(f)$$
$$h(t) = s(t)$$

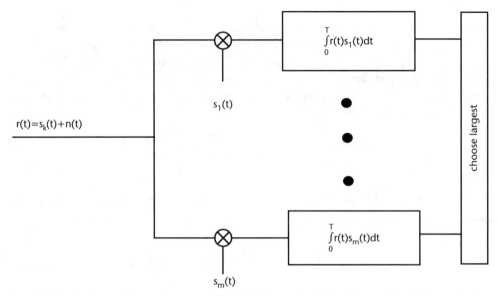

Figure 5.4 Ideal correlation receiver architecture. The received signal is correlated with each of the possible message waveforms. The correlation with the largest value is chosen as the transmitted signal.

The filter is matched to the signal, or more commonly put, this is a matched filter. As a simple example, consider the NRZ pulse.

The output of the matched filter at the sampling time T is now

$$y(t) = \int r(\tau)h(t-\tau)d\tau$$

$$y(T) = \int_0^T r(t)h(t)dt$$

This result is exactly the same as that for the correlation receiver. Figure 5.6 is the block diagram of an optimum matched filter receiver.

Consider this important question: What is the SNR measured at the sample time of the matched filter? We need to develop $y(T)$ further to get the answer:

What is the form of h(t) which maximizes the output SNR at some time T?

h(t) = s(T-t)

This is called the matched filter.

Figure 5.5 Optimum filter maximizes output SNR at sample time T.

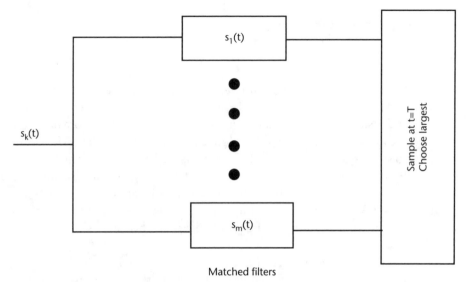

Figure 5.6   Matched filter receiver. The received waveform is passed through matched filters representing the set of possible signals. At the sample time T, the results are compared and the largest value is chosen as the transmitted signal.

$$y = \int_0^T r(t)h(t)$$
$$= \int_0^T \left[s(t) + n(t) \right] h(t) dt$$
$$= \int_0^T s(t)s(t)dt + \int_0^T n(t)h(t)dt$$
$$s(t) = h(t) \quad \text{(matched filter)}$$
$$E = \int_0^T s^2(t)dt$$

The quantity E is the signal energy. Now the SNR is defined as

$$SNR = \overline{y^2} \Big/ \overline{(y - \overline{y})}^2$$

Since the noise is zero mean we immediately have $\overline{y^2}(t) = E$. The noise term takes a little more doing:

$$\overline{(y - \overline{y})}^2 = \overline{\int_0^T h(t)n(t)dt \int_0^T h(t')n(t')dt'}$$
$$= \int_0^T \int_0^T h(t)h(t')\overline{n(t)n(t')}dtdt'$$
$$= \int_0^T \int_0^T h(t)h(t')_0 \left[\frac{N_0}{2}\delta(t - t') \right] dtdt'$$
$$= \frac{N_0}{2} \int_0^T h^2(t)dt = \frac{N_0 E}{2}$$

Combining the two pieces gives

$$SNR = 2\,E/N_0$$

This result is remarkably simple and powerful at the same time. The maximum output SNR of a matched filter depends only on the signal energy E and the noise power density N_0. Observe that there is no statement here as to what the signal looks like in time or what the signal PSD looks like in frequency. This situation comes into play in some systems where the SNR is measured as the power in some bandwidth B, usually the IF bandwidth of the receiver. Changing B changes this measure of the SNR, but not the basic result $2E/N_0$. In short, you cannot cheat the definition to suit a purpose! Another fact is that any other filter inserted between the transmitted signal and the matched filter will degrade the SNR, however slightly.

The Integrate and Dump (I&D) Matched Filter

The I&D is a common matched filter. Consider the NRZ pulse

$$P(t) = 1 \quad 0 \le t \le T$$

The Laplace transform of this pulse, which is also the transform of its matched filter, is given by

$$H(s) = \left(1 - e^{-Ts}\right)/s$$

The matched filter is realized by sampling the output at times 0, T, $2T$, etc.

Where does the term I&D come from? To answer this, consider the simple RC circuit shown in Figure 5.7. From circuit theory, the transfer function from the input to the capacitor voltage is given by

$$v_c(s)/v_{in}(s) = 1/(s + 1/\tau)$$

$\tau = RC$ is the circuit time constant. When τ is large, the transfer function acts simply as an integrator. The voltage builds up on the capacitor. But what happens if we momentarily short out the capacitor every T seconds? Then the voltage goes to zero and the integration starts over again; I&D. The symbol $\int\downarrow$ is commonly used to demote this operation. Figure 5.8 shows a block diagram of the equivalent circuit in the time domain. It solves the differential equation

$$R\,dq/dt + q/C = V(t)$$

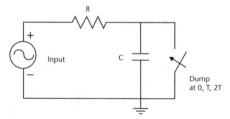

Figure 5.7   RC integrate and dump matched filter.

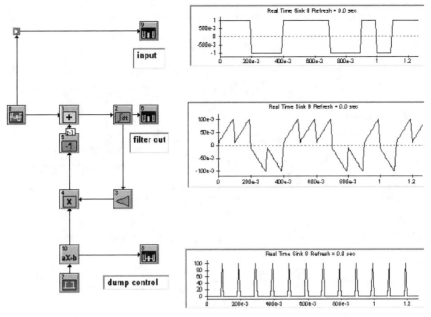

Figure 5.8   True integrate and dump matched filter simulation.

In this model we control the effective capacitance by the square wave time function.

5.4   From Signals to Vectors

We started this chapter with a discussion on a mathematical vector receiver. We followed that with the optimum detection of a series of time-based waveforms. If all we need is the latter discussion, why introduce the vector channel in the first place? We take this issue up now.

First of all, note that we can make a set of time basis functions as a basis for some vector analysis. The handbooks are full of sets of orthogonal functions that can be used for various purposes. What we require is a set of time functions $\{s_k(t)\}$ having the properties

$$\int_0^T s_k(t)s_p(t)dt = 0 \quad k \neq p$$

$$= 1 \quad k = p$$

Such a set is said to be orthonormal. An arbitrary function $s(t)$ can then be written in the expanded form

$$s(t) = \sum_{k=0}^{N-1} c_k s_k(t)$$

$$c_k = \int_0^T s(t)s_k(t)dt$$

The waveform is completely described by the set of coefficients $\{c_k\}$. So at the transmitter we choose the particular members of this set to describe the message waveform of interest. At the receiver, the signal is corrupted by noise $n(t)$ presenting a waveform $r(t) = s(t) + n(t)$. The objective of the receiver is to make the best estimate of $\{c_k\}$, denoted as $\{\overline{c}_k\}$. The following shows the steps to accomplish this:

$$r(t) = s(t) + n(t)$$

$$= \sum_{k=0}^{N-1} c_k s_k(t) + n(t)$$

$$\overline{c}_k = \int_0^T r(t) s_k(t) dt$$

$$= \int_0^T \left[\sum_{k=0}^{N-1} c_k s_k(t) + n(t) \right] s_k(t) dt$$

$$= c_k + n_k$$

$$n_k = \int_0^T n(t) s_k(t) dt$$

What we now have is a received vector set $\{\overline{c}_k\}$, which is the sum of the true vector set $\{c_k\}$ plus statistically independent noise vector set $\{n_k\}$. The bottom line here is that we have reduced the signal problem back to the previously described vector problem, with all of the rules still in force.

An important illustration of the error probability in a two-dimensional case is the comparison of two basic waveforms:

- *Binary FSK:* In this method a logical [1] is sent as some frequency f_1, and a logical [0] is sent as another frequency f_2.

$$s_1(t) = \cos[2\pi f_1 t]$$
$$s_2(t) = \cos[2\pi f_2 t]$$
$$0 \le t \le T$$

where T is the signaling time at rate $r = 1/T$. We choose the frequencies such that

$$\int_0^T s_1(t) s_2(t) dt = 0$$

This can be satisfied if $f_1 = n/T, f_2 = m/T$. In this case the two waveforms are the basis vectors of a two-dimensional space, so we can write

$$s_1(t) \leftrightarrow [0,1]$$
$$s_2(t) \leftrightarrow [1,0]$$

- *Binary PSK:* In this mode, a logical [1] is transmitted as a 0 phase with respect to the carrier, and the logical [0] is transmitted as a relative phase of π. The two waveforms are then (remembering that $\cos(\pi) = -1$)

$$s_1(t) = \cos(2\pi f_0 t)$$
$$s_2(t) = -\cos(2\pi f_0 t)$$

such that

$$\int_0^T s_1(t)s_2(t)dt = -1$$

From this example the vector relation relative to a two-dimensional space is

$$s_1(t) \leftrightarrow [1,0]$$
$$s_2(t) \leftrightarrow [-1,0]$$

The two cases are combined and shown in Figure 5.9.

What is the difference in BER performance of these two cases? The answer was already derived at the beginning of the chapter. The BPSK is clearly the same scalar case since it only has one vector component. The FSK can also be reduced to the same type of scalar case by a rotation and translation of the axis, shown in Figure 5.8. In either case, the BER is dependant only on the distance, d, between the signal vectors giving the results

$$P(e) = Q\left[\sqrt{2d^2/\sigma^2}\right] = Q[2E_b/N_0] \quad \text{antipodal}$$
$$P(e) = Q\left[\sqrt{d^2/\sigma^2}\right] = Q[E_b/N_0] \quad \text{orthogonal}$$

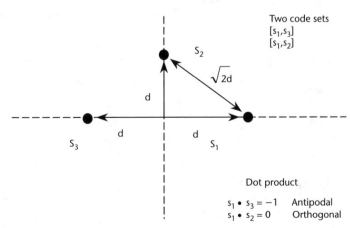

Figure 5.9 Two-dimensional vector representation of orthogonal FSK, and antipodal BPSK. The BPSK's distance squared is twice that for FSK.

Figure 5.10 plots these two BER curves relative to the E_b/N_0. correlation receiver. Of course, both cases must go to BER = 0.5 as $E_b/N_0 \to 0$; but observe the difference as $E_b/N_0 \to 8$. The BPSK (antipodal) case has a 3-dB advantage over FSK (orthogonal) for the same BER. In other words, it takes half the signal energy (power) for the same performance. It is useful to remember the indicated point on the plot: BPSK BER = 10^{-5} for an E_b/N_0 of 9.5 dB.

5.5 The I, Q Vector Space

The correlation receiver, matched filter receiver, and the vector receiver all produce theoretically optimum procedures for recovering the signal in noise. The problem is that the computational load to implement them might be very high. This is especially true when considering higher order modulations such as Mary phaseshift keying (MPSK) and quadrature amplitude modulation (QAM). Consider a not so unusual situation as 256 QAM modulation. We could implement the optimum detector with 256 correlators. This would be very time consuming. What we want is a minimum basis set. One possibility is the set used in Fourier series

$$\{s_k(t)\} = \{\sin(2\pi k/T), \cos(2\pi k/T)\}$$

One would use this set for MFSK modulation. However, as the number of messages, N, increases, the basis set increases in proportion. The receiver would still be required to perform N correlations. We stated before that MFSK is not generally used because it increases the signal bandwidth, and add to this the current fact about the vector basis size limits MFSK to a maximum of four, for most practical purposes. But observe that the functions $\sin(2\pi t/T)$, $\cos(2\pi t/T)$ form a two-dimensional orthogonal basis vector set since

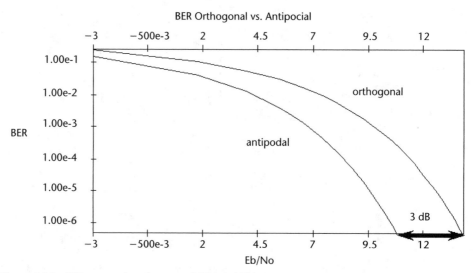

Figure 5.10   BER comparison between BPSK and FSK.

$$\int_0^T \cos(2\pi t/T)\sin(2\pi t/T)dt = 0$$

Thus a large majority of current systems are based on multiple vectors in a two-dimensional space of the form

$$s(t) = I(t)\cos(2\pi ft) + Q(t)\sin(2\pi ft)$$

The I signal is commonly called the in-phase component, and the Q signal is the quadrature component. For quadra phase shift keying (QPSK), I = ±1 and Q= ±1 also. Furthermore, note that the physics of the transmitting sine wave signals $s(t)$ is generally in the form

$$
\begin{aligned}
s(t) &= A(t)\sin\big[\varphi(t)\big] \\
&= A(t)\sin\big[2\pi f_0 t + \theta(t)\big] \\
&= A(t)\big[\sin(2\pi f_0 t)\cos(\theta(t)) + \cos(2\pi f_0 t)\sin(\theta(t))\big] \\
&= I(t)\cos(2\pi f_0 t) + Q(t)\sin(2\pi f_0 t)
\end{aligned}
$$

This is true even for MFSK signals.

This I,Q representation is the basis for virtually all communication systems.

5.6   Conclusion

In this chapter we developed the basic methodology for optimum detection of digital signals. We started with the concept of vector signals, and showed that the optimum detector chooses the output that is the nearest distance to a possible member of the vector set. We then went from vectors to time functions. First we derived two forms of an optimum waveform receiver, namely, the correlation and matched filter receiver. Next we showed how to represent a time waveform as a linear combination of an orthogonal based set. This representation then provided the connection between the time waveform detection and the vector space concept. As a result of this, we showed that antipodal vectors (BPSK) enjoyed a 3-dB detection advantage over orthogonal vectors (FSK). Finally we observed that the most commonly used structure was the two-dimensional IQ format.

Selected Bibliography

Pahlavan, K., and A. H. Leveseque, *Wireless Information Networks,* New York: John Wiley & Sons, 1995.

Proakis, J. B., *Digital Communications,* New York: McGraw-Hill 1983.

Rappaport, T. S., *Wireless Communications, Principles and Practice,* Upper Saddle River, NJ: Prentice Hall, 2002.

Sklar, B., *Digital Communications, Fundamentals and Applications,* Upper Saddle River, NJ: Prentice Hall, 2001.

Steele, R., *Mobile Radio Communications,* London: Pentech Press Limited, 1992.

CHAPTER 6

Modulation

Once the transmitter has produced the final bit stream, the last step is to modulate it onto a suitable carrier. This chapter describes the most commonly used modulations. It is possible to vary only the amplitude, phase, or frequency of a sine wave. So all modulation formats are some theme and variation of using these three possibilities. In general, simulating a modulator is a relatively easy task, much easier than that of the receiver.

Our emphasis in this chapter will be on digital modulation as opposed to standard AM and FM modulation.

6.1 Amplitude Modulation (MASK)

One of the simplest forms of digital modulation is multiple amplitude shift keying (MASK). In its simplest form, a logical [1] is transmitted with some amplitude A, and a logical [0] is not transmitted at all. This format is commonly called on-off keying (OOK). Figure 6.1 shows an input NRZ data, and Figure 6.2 shows the resulting transmitted signal on a carrier. Figure 6.3 shows the same idea with four levels of amplitude, 4ASK, Figure 6.4 shows the corresponding modulated signal.

We can continue this process indefinitely. Take k bits in a group to make one of 2^k amplitudes to be modulated. As k becomes large, the required bandwidth becomes smaller, but the decision levels required to differentiate the levels become closer together, thus causing a loss or performance.

6.2 Frequency Modulation (MFSK)

In two-tone multiple frequency shift keying (MFSK) a logical [1] is transmitted at some frequency f_1, and a logical [0] is transmitted on some other frequency f_2 (see Figure 6.5).

We now FSK modulate this data pattern using two tones, 1 and 2 Hz. The resulting modulated waveform is shown in Figure 6.6.

The PSD of this modulation is shown in Figure 6.7 Note that there are two unmodulated (CW) components at 1 and 2 Hz. These CW tones carry half of the modulated signal energy. This is wasted energy as far as detection is concerned. This is why the FSK example of Chapter 5 shows a 3-dB loss in detection performance.

Clearly, this idea can be extended to 4FSK, 8FSK, and so on. But take a look at the evolution of the signal PSD as we do this. If the original data rate is $R = 1/T$, then every T seconds we send one of two frequencies separated by R Hz. The bandwidth

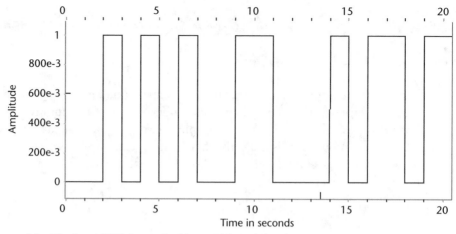

Figure 6.1   The input NRZ data to be binary ASK modulated.

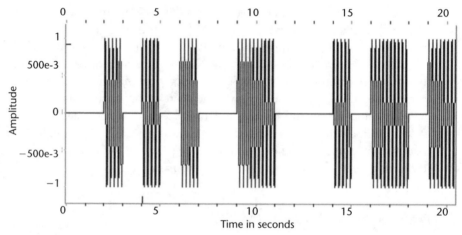

Figure 6.2   Modulated ASK signal corresponding to the data of Figure 6.1.

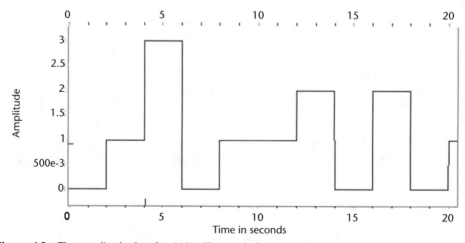

Figure 6.3   The amplitude data for 4ASK. The symbol rate is half the bit rate.

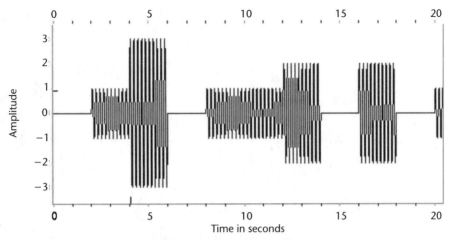

Figure 6.4   MASK modulated signal from the data of Figure 6.3.

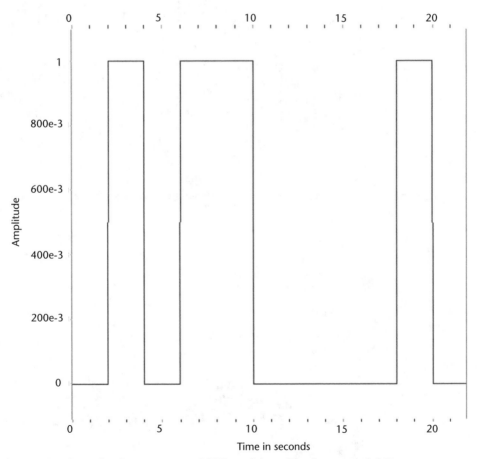

Figure 6.5   Binary data into a two-tone MFSK modulator. The data rate is 0.5 Hz.

is then about $2R$ as seen in Figure 6.7. Now take 3 bits at a time. Then we transmit one of $2^3 = 8$ Frequencies separated by $8/3T$ seconds for a bandwidth of $4R$. One more time: Taking 4 bits, we send one of 16 frequencies every $1/4T$ seconds requir-

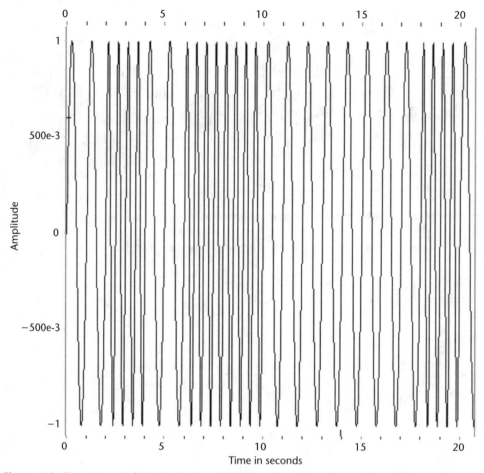

Figure 6.6   Two-tone modulated signal resulting from the data pattern of Figure 6.5.

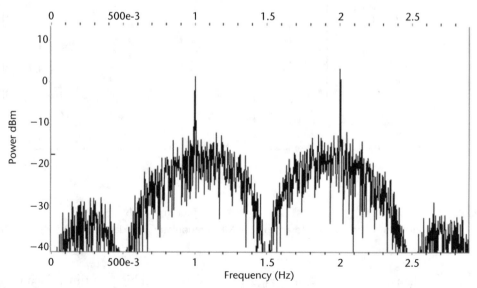

Figure 6.7   PSD of two-tone FSK. Note the CW spike components at the two modulated frequencies, 1 and 2 Hz.

ing a bandwidth of 16/4T. In general for MFSK the bandwidth increases as $2^M/M$, which is not a good thing. For this reason MFSK is not commonly used; the highest seen by the author is 4FSK.

6.3 Phase Modulation (MPSK)

Multiple phase shift keying (MPSK) modulation is the most commonly employed format used due to the relative simplicity of the modulator and demodulator. The simplest version is BPSK, where a logical 1 is encoded as 0 phase, and a logical 1 is coded as a phase of π. This is represented by the equation

$$s(t) = A\sin(2\pi f_0 t + a_k \pi) \quad kT \le t \le (k+1)T$$
$$= Aa_k \sin(2\pi f_0 t) \quad a_k = \pm 1$$

This format can be extended to higher orders

$$s(t) = A\sin(2\pi f_0 t + \varphi_k)$$

where the phase set for QPSK is given by

$$\varphi_k = 2^k \pi/4 \quad k = 0,1,2,3$$

This general idea can be extended to 8PSK, which encodes 3 bits of data into one of eight phases, and so forth.

At this point we introduce the useful concept of a signal constellation. The idea is to plot the possible values of the I signal against the corresponding values of the Q signal. This is illustrated in Figure 6.8 for 8PSK. Note that the constellation is simply eight points (phases) on a unit circle since for this modulation we have $I^2 + Q^2 =$ constant = power. Furthermore, this figure shows two methods of encoding the data

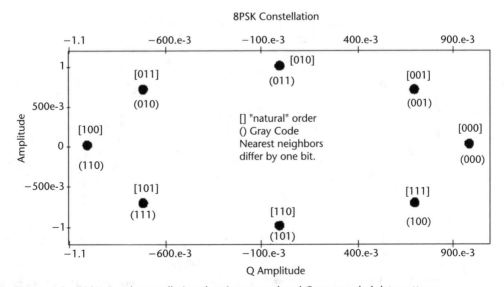

Figure 6.8 8PSK signal constellation showing normal and Gray encoded data patterns.

into those phases. In the first method, known as natural order [], the encoder simply maps the phase into the binary count order of the bits as shown: [000] = 0, [001] = $\pi/4$, and so forth, around the circle. But look closely at the encoding [111] into $7\pi/4$. As can be seen, this symbol encoding is the nearest constellation member to the [000] code word phase. From detection theory, if the symbol [000] is transmitted, then the most likely error is the two nearest neighbors in the constellation, $7\pi/4$ being one of them. In this case the decoded symbol will produce 3 bit errors. The solution is called Gray encoding and is illustrated by the assignment () in Figure 6.8. Now the error caused by nearest neighbor decoding is only 1 bit of 3 no matter where in the constellation the original signal is.

Why go to higher order modulation? The answer is bandwidth. The group of k bits into one of the 2^k phases is called a symbol, as opposed to the term bit. Suppose that the original data rate is 1 bps. Then a QPSK symbol would change every 2 bits or at a 0.5 symbols/sec rate. The 8PSK would encode into symbols at a rate of 1/3 symbols/sec, and so forth. The occupied spectra of these cases are shown in Figure 6.9.

As Figure 6.9 shows, the higher the order, the more compact the occupied spectra, which is a good thing.

So if 8PSK does better in bandwidth than QPSK, why not continue this process to a much higher order PSK with the attendant savings in bandwidth? Unfortunately, Mother Nature does not cooperate in another overriding area: detection. From detection theory, the factor that sets the system BER is the distance between the nearest constellation points. For a given bit rate for MPSK, as the order increases, the diameter of the constellation increases since the symbol energy

$$E_s = A^2\,T/2 = A^2 kT_b/2 = kE_b$$

is increasing. At the same time the number of points on the circle increases as well. The result of this process is shown in Figure 6.10. The general expression for this distance, d, is given by

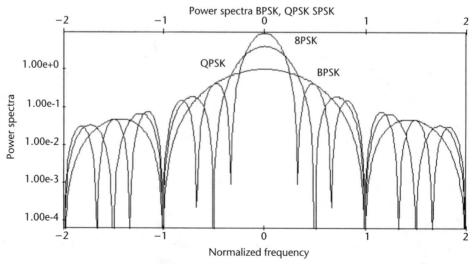

Figure 6.9 PSD of BPSK, QPSK, and 8PSK modulation. As the modulation order increases, the occupied signal bandwidth decreases.

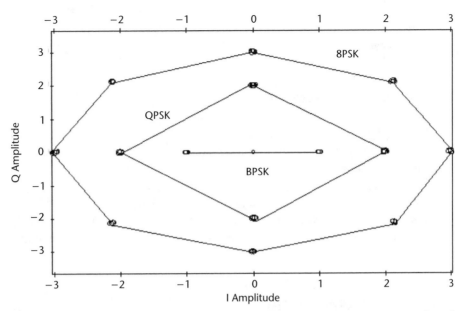

Figure 6.10 Vector signal constellations for MPSK. The diameter of the circle increases, but after QPSK the number of points on the circle increases at a faster rate, thus reducing the distance between the nearest neighbor points.

$$d_k^2 = 4kE_b \sin^2\left(\pi/2^k\right)$$

which is plotted in Figure 6.11. From this figure we observe *a very important observation*. The distance for BPSK is the *same* as the distance for QPSK. Therefore, the BER performance for the two is the same. But we have also shown that the QPSK only requires half the bandwidth of the PSK. From Figure 6.11, we also see that the distance (performance) decreases steadily beyond QPSK. This fact is why QPSK is the workhorse modulation type of nearly all modern digital communication sys-

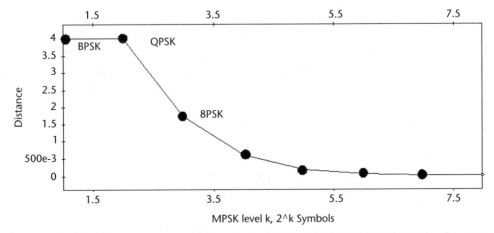

Figure 6.11 The distance between constellation points for MPSK. BPSK and QPSK have the same distance. Beyond that, the distance and performance decreases.

tems. Some 8PSK systems do exist, such as GSM/EDGE. It is nearly impossible to find any modulation order higher than 8PSK in any modern system.

6.4 $\pi/4$ DQPSK Modulation

$\pi/4$ DQPSK modulation is used in many wireless formats such as IS-136. The input data bits in groups of two are encoded to phase according to the differential rule shown in Table 6.1.

The signal constellation for this format is shown in Figure 6.12. Observe that the transition from one state to another never goes through the origin of the graph. This is important to the hardware designers. For efficient use of amplifiers, it is desired to minimize the peak to average ratio of the signal. This is important to the hardware designers.

6.5 Offset QPSK Modulation

Offset QPSK modulation is similar to QPSK, except that one of the data channels, Q, is delayed by 1/2 of a bit time with respect to the I channel. This is modeled by the equation

$$s(t) = I(t)\cos(2\pi f_0 t) + Q(t - T/2)\cos(2\pi f_0 t)$$

It can be seen that the signal phase can only change by $\pm\pi/2$ every $T/2$ seconds. The signal constellation is shown in Figure 6.13.

As with the $\pi/4$ DQPSK, there is no transition path from one state to another that passes through the origin.

Offset QPSK is used by the IS-95 wireless system for the return (mobile to base) link.

6.6 QAM Modulation

We saw that for MPSK the signal distance decreases beyond QPSK. Note in Figure 6.10 also that the constellation has a huge hole in it where one might place signals that have a greater distance. This is the effect when quadrature amplitude modula-

Table 6.1 PI/4 DaPSK Symbol to Symbol Phase Change

Symbol	Phase Transition (deg)
00	45
01	135
10	−45
00	−135

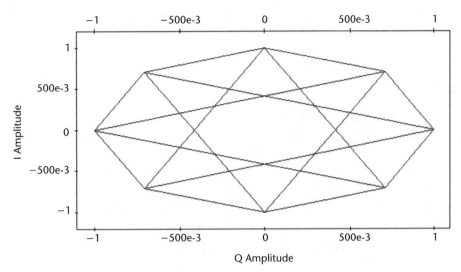

Figure 6.12   Constellation of π/4 DQPSK modulation. No state transition goes through the origin.

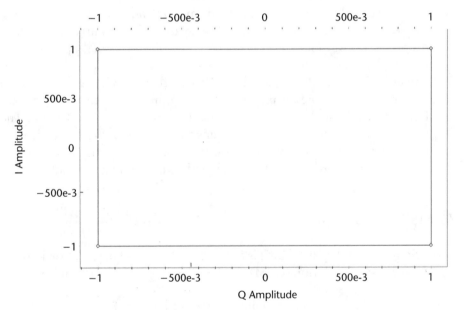

Figure 6.13   Constellation of offset QPSK.

tion (QAM) is employed. QAM is similar to ASK modulation except that we put an ASK signal in the in-phase (cosine) and quadrature (sine) modulation term. The mathematical description of this signal is

$$s(t) = a_k \cos(2\pi f_0 t) + b_k \sin(2\pi f_0 t)$$

where a_k and b_k are multiamplitude values of the form $[\pm 1, \pm 3, \pm 5, ..., \pm 2M{-}1]$. For $M = 2$, there are four terms on the I component and four terms on the Q component. Figure 6.14 shows a constellation in this case.

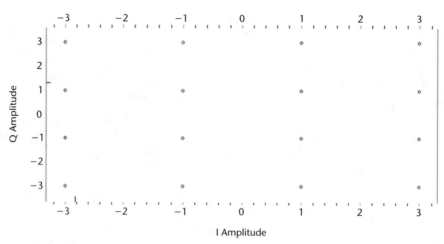

Figure 6.14 Constellation of 16QAM modulation.

There are 16 points in the figure, and this is denoted by 16QAM. With eight elements on a side, we have 64QAM and so forth. More complicated configurations are also possible that are not rectangular, and are not based on a 2^L basis. Just like MPSK, as the order of the QAM increases, the required transmit bandwidth decreases. QAM modulation is the basis of the various dial-up modems commonly used in computer to Internet communications. QAM orders of 256 and greater are common. The price that is paid, like MPSK, is that the BER performance degrades for higher orders. But the telephone channel is a high SNR entity, so the performance loss can be tolerated while gaining valuable bandwidth compression. The common telephone has a bandpass filter that runs from 300 to 3,000 Hz, which is the spectra occupancy of the human voice. Thus, to increase the bit rate to the max, 56K at the moment, a high order of QAM along with exotic coding techniques is required.

6.7 MSK Modulation

Minimum shift keying (MSK) is a member of a class of modulation known as continuous phase modulation (CPM). MSK is derived from the formula

$$s(t) = A \sin\left(2\pi f_0 t + 2\pi\mu \int m(t) dt\right)$$

where $m(t)$ is the NRZ data waveform with rate $R = 1/T$. The modulation gain μ is chosen such that the signal phase advances by $\pm\pi/2$ over a bit period T. This implies that

$$2\pi\mu \int_0^T (1) dt = \pi/2$$

$$\mu = R/4$$

The advantage of MSK is that the spectral occupancy is much more compact than that of the BPSK signal of the same rate. In a BPSK signal the phase can instantly jump from 0 to π at the bit transition. This jump causes the wide occupied spectrum of the BPSK signal. In MSK the phase is continuous across the bit transition owing to the integral phase representation.

A common variant of MSK is called GMSK (the G stands for Gaussian). Figure 6.15 shows this technique. It shows that the input data $m(t)$ is passed through a Gaussian low pass filter whose 3-dB bandwidth B is determined by $BT = 0.3$. GMSK is the modulation of the GSM wireless format used extensively in Europe and around the world. The Gaussian filter is used to further compact the signal spectrum as shown in Figure 6.16.

6.8   OFDM Modulation

Orthogonal frequency division modulation is used in many wireless systems such as the 802.11g standard. Another name is multicarrier modulation (MCM). In Chapter 10 on Channel Models we describe the effect of a single delay path that combines with the direct signal. The effect of this is to put notches in the transmitted signal spectra as seen at the receiver. In a complicated environment, there could be several such notches, which can come and go as the system dynamics change. One approach to this problem is to employ an equalizer that some how knows or learns what the channel is doing, and then makes the appropriate correction.

In the OFDM approach, instead of modulating all of the high speed data onto a single carrier, the trick is to modulate many narrowband carriers each bearing part of the original data stream. For example, a 100-Mbps channel can be QPSK modulated on a carrier with a null-to-null bandwidth of 200 MHz. An alternative might be to split the data into five data streams with data rate 20 Mbps, and modulate each on a carrier separated by 40 MHz. This is illustrated in Figure 6.17(a–d). In reality, the number of such carriers can be 1,000 or more. The 802.11g wireless LAN standard uses some 1,700 carriers.

What is gained by this format? Return to the frequency selective fade channel. Now the fade notches will only eliminate a small number of the carriers specifically where they occur. By proper FEC and interleaving, the receiver can process through these notches.

In complex notation we can write an OFDM modulation in the form

$$s(t) = \sum_{k=0}^{N-1} c_k e^{-2\pi jkt/T}$$

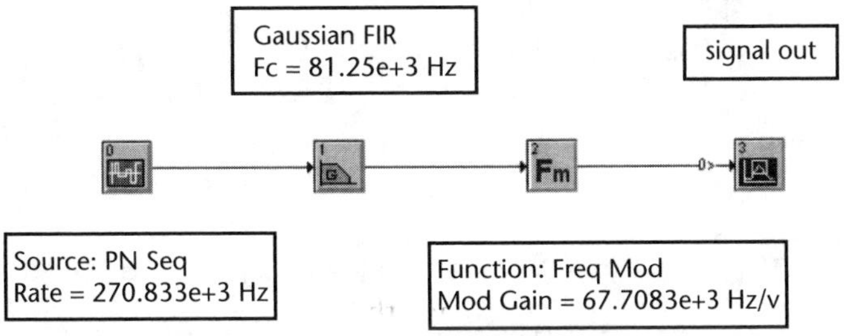

Figure 6.15   Block diagram of GMSK modulator.

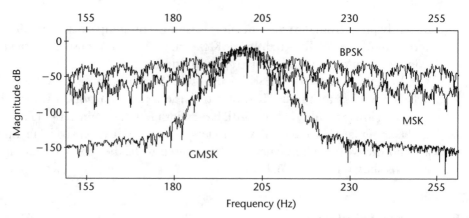

Figure 6.16 PSD of BPSK, MSK, and GMSK modulation. Observe the dramatic effect of the Gaussian filter in GMSK.

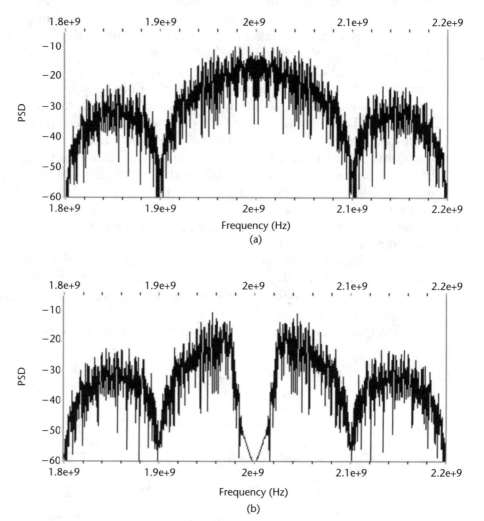

Figure 6.17 (a) Standard PSD of a PSK modulated signal with a 100-MHZ data rate. (b) PSD of the 100-MHz signal of (a) with a frequency selective fade notch. (c) PSD of five separate PSK signals with 20-MHz data. (d) PSD of the five-carrier system of (c) with a frequency selective fade notch. Note, only the one central carrier is affected.

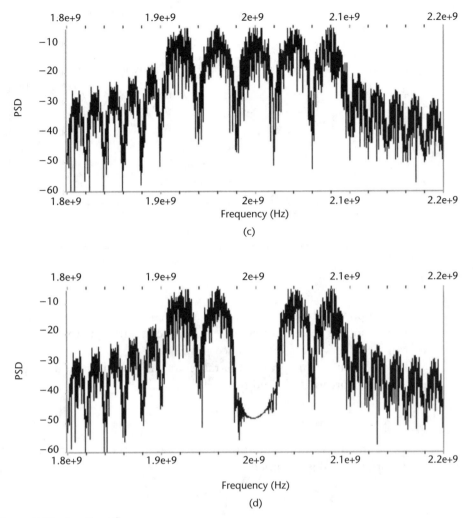

Figure 6.17   (continued).

where c_k is the complex [I, Q] symbol modulating the kth carrier, and $1/T$ is the carrier spacing. The above looks something like the DFT algorithm described in Chapter 2. In fact, the generation algorithm generally employs an FFT even if N is not some nice power of 2. What is done is to increase the actual N to a power of 2, and simply tell the FFT that the coefficients c_k for the appended noncarriers are simply zero.

There is one final and very important feature of OFDM. In a frequency selective fade environment, the delay at one frequency will be different than the delay at another. This is illustrated in Figure 6.18 where the symbol time is 1 second, and there are four carriers. Figure 6.18 illustrates what happens to one such OFDM symbol at the receiver. The vertical bars are aligned with the 4-Hz component as a reference for the relative delays of the other tones.

What we see is that the timing of the various components of the OFDM signal are shifted with respect to each other. Now what happens if we try to use the inverse FFT operation for the demodulator? For this to work, all of the individual fre-

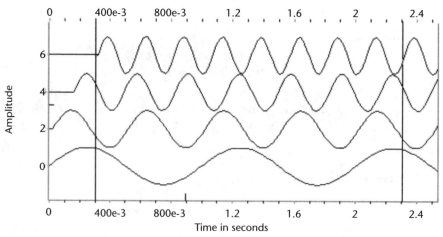

Figure 6.18 Four OFDM carrier symbols each with a different time delay due to multipath. The start and stop time of each are not aligned as required by the demodulator.

quency components must line up in time at the receiver, as they were at the transmitter. This is not the case in the stated environment. Attempts to use the inverse FFT will run into intersymbol interference (ISI).

The solution to this problem is rather ingenious. First we denote the already defined delay spread as T_Δ, and the useful time as

$$T_u = T - T_\Delta$$

At the modulator, the symbols are still presented at T second interval. But the FFT algorithm used is modified to

$$s(t) = \sum_{k=0}^{N-1} c_k e^{-2\pi kt/T_u}$$

In other words, the orthogonal basis in time is T_u not T. The time associated with the above is to produce a symbol of $T_u < T$ seconds long. The trick is to extend $s(t)$ as generated by the FFT from T_u to T seconds.

At the receiver, as stated, different frequency components encounter different delays. But as long as the delay spread is less than T_u, there is always a segment within the T seconds of the symbol of T_u seconds long that does not suffer any overlap. The receiver simply truncates the appropriate portion of the symbol down to T_u, and performs the inverse FFT recovering the data.

The 802.11g wireless LAN specification uses $T = 4$ usec, $T_u = 3.2$ usec, and $T_d = 0.8$ usec.

6.9 Pulse Position Modulation

In pulse position modulation (PPM), the information amplitude is converted into a pulse whose position with respect to the symbol time marker is proportional to that amplitude. PPM is one format that has been proposed for UWB modulation systems.

Figure 6.19 shows the representative input set of data signals, and Figure 6.20 shows the resulting PPM modulation. In this case the input symbols are at a rate of 1 Hz. The lowest symbol at 0, corresponding to the bit pair [0,0], is mapped to a zero pulse position offset from the frame boundary. The +1 amplitude symbol, corresponding to the bit pair [0,1], is displaced by 1/4 of the symbol time, and so on.

While this example shows 4-PPM, it can be extended further to 8-PPM, and beyond.

6.10   Pulse Width Modulation

In pulse width modulation (PWM), the width of the pulse with respect to the symbol time marker is proportional to the data. Figure 6.21 shows the input four level symbols as in the PPM case. Figure 6.22 shows that the width of the pulse from the symbol boundary is proportional to that information. In Figure 6.22 the pulses were given a Gaussian shape to make the presentation clearer.

As with the PPM case, this idea can be extended to higher order modulation.

6.11   GSM EDGE Modulation

The Global System for Mobile Communication (GSM) Enhanced Data Rate for Global Evolution (EDGE) system was developed to provide high data rate to the GMSK wireless system. The modulation is 8PSK with a twist. Figure 6.23 shows the basic block diagram.

The top row of functions is a standard 8PSK modulator. The input bits are converted in groups of three to one of eight symbols [0, 7]. Each symbol is then modulated as one of eight phases. However, the phases generated at this point are all advanced by a rotating factor of 3/8 radians. To see the effect of this rotation, Figure 6.24 shows the signal constellation obtained.

The important feature here is that the transitions from one state to another never go through the origin of the diagram. This is the same feature as noticed for the IS-136, $\pi/4$ DQPSK system.

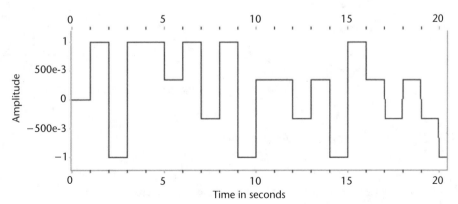

Figure 6.19   Input symbol stream. The larger the amplitude, the greater the pulse position from the symbol boundary.

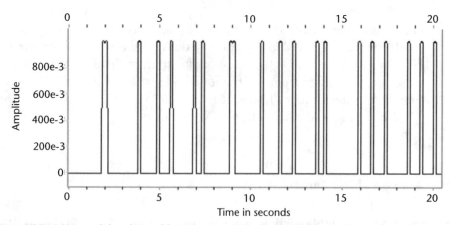

Figure 6.20   PPM modulated signal based on symbols shown in Figure 6.19.

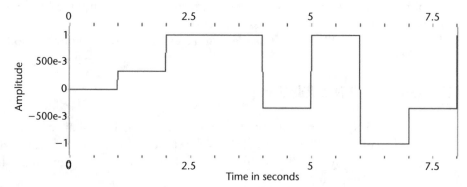

Figure 6.21   Input four level symbols into a PWM modulator.

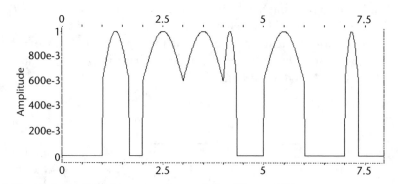

Figure 6.22   Modulated PWM based on the symbols of Figure 6.21.

6.12   Spread Spectrum Modulation

Spread spectrum communications were originally developed for military purposes, but now they have a wide variety of uses in the public sector as well. The world of spread spectrum is divided into two general categories: frequency hopping and direct sequence. Some applications include the following:

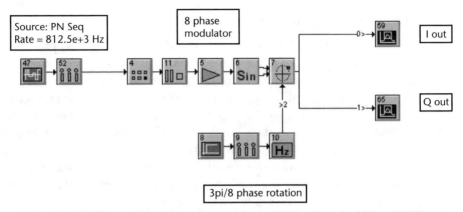

Figure 6.23 GSM EDGE modulator. The system is a basic 8-PSK with and additional 3 ð/8 rotation on each symbol.

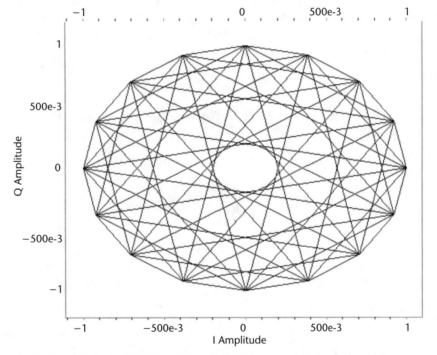

Figure 6.24 Constellation of GSM EDGE signal. There is no path through the origin.

- Antijam (AJ);
- Low probability of intercept;
- Ranging;
- Antimultipath;
- Multiple access (e.g., CDMA);
- Antiinterference.

(See Appendix D for an interesting trivia question relating to spread spectrum technology.)

6.12.1 Frequency Hopping

Frequency hopping (FH) was originated and is still used for AJ purposes. U.S. Army tactical radios operating in the 30 to 88-MHz band use this technology. The idea is quite simple. If the jammer jumps on your frequency, move to another one. The whole process is coordinated where frequency hop occurs at regular intervals, the hop rate. The sequence of the hops is known to both the transmitter and receiver (but not the jammer!) so they stay in step. The jammer must either guess the next hop, or broaden the bandwidth of his signal to try and cover all possibilities at once. For a fixed power jammer, this broadening reduces the delivered jamming power per channel. Figure 6.25 shows the basic block diagram of an FH system. Figure 6.26 shows the concept of the FH in time.

6.12.2 Direct Sequence Pseudo Noise

The direct sequence pseudo noise (DSPN) system works as follows. The basic NRZ data wave form $d(t)$ of rate R_d, is multiplied by the logic exclusive OR (XOR) with a second spreading code $p_n(t)$ that has a chip rate $R_p \gg R_d$. The transmitted signal is simply

$$s(t) = d(t)p_n(t)$$

At the receiver (once synchronized) the signal is again multiplied by $p_n(t)$. Under the proviso that $p_n^2(t) = 1$, the net result is

$$r(t) = s(t)p_n(t)$$
$$= d(t)p_n(t) \cdot p_n(t)$$
$$= d(t)$$

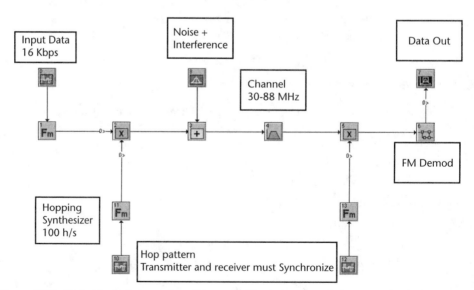

Figure 6.25 Basic block diagram of an FH system. The hop pattern is known to both the transmitter and receiver.

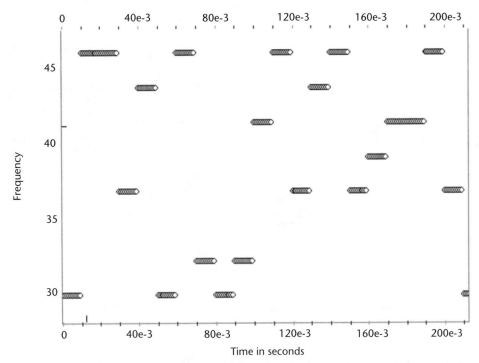

Figure 6.26 FH hop pattern. Every *T* seconds, the transmitter moves from one frequency to the next.

So we have taken a few extra steps in the modulation process to spread and despread the signal. The process of multiplying by $p_n(t)$ is called spectrum spreading. The reverse process of recovering $d(t)$ is called spectrum collapsing or despreading. In short, we go from a narrowband signal to a wideband signal and back again to the original narrowband signal.

Figure 6.27 shows a basic block diagram of the process. Here we insert all of the bad things that can happen to a signal between the transmitter and receiver. Let us take these impairments one at a time and see what we have gained by this operation. It should be significant since we have increased the required bandwidth of the sigal, which is usually a bad thing.

The receiver applies $p_n(t)$ to the total received signal $r(t)$:

$$r(t) = d(t)p_n(t) + nb(t) + wb(t) + mp(t) + n(t)$$

where $nb(t)$ is narrowband interference, $wb(t)$ is wideband interference, $mp(t)$ is the multipath signal, and $n(t)$ is additive white noise.

Now we multiply the received signal by the dispreading code $p_n(t)$ to get

$$r(t)p_n(t) = d(t) + nb(t)p_n(t) + mp(t)p_n(t) + n(t)p_n(t)$$

As said, $d(t)$ is narrowband, but what about the rest of the terms? It turns out they are all wideband as long as the multipath delays are greater than one chip of the *pn* code.

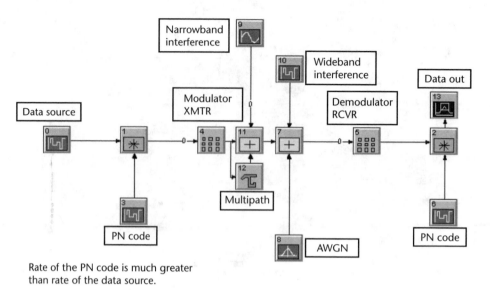

Rate of the PN code is much greater
than rate of the data source.

Figure 6.27   Basic block diagram of an DSPN system showing the possible impairments added between the transmitter and receiver.

Figure 6.28 (a–f) illustrates these concepts for a case with a data rate of 10 Kbps spread by a $p_n(t)$ code of rate 1 Mbps. There is a CW interferer at 500 kHz from the center of the modulated signal. Figure 6.28(a) shows the input data stream. Figure 6.28(b) shows the PSD of the input data stream. The nulls in the spectra are at multiples of 10 kHz. Figure 6.28(c) shows the PSD of the $p_n(t)$ spread signal with the CW interference. The nulls in the spectra are at multiples of 1-MHz spreading code rate. At the receiver the signal is multiplied by the same pn sequence as used in the modulation. Figure 6.28(d) shows the resulting PSD after this operation. Note that the $p_n(t)$ signal is still there, but it is now centered on the location of the interference. The centered narrowband PSD is the collapsed spectra of the original data. Figure 6.28(e) shows a zoom of this portion of the spectra. Figure 6.28(f) shows the recovered data signal after narrowband filtering the signal of Figure 6.28(d) around the data only. This eliminates most of the residual spread signal that is now centered on 500 kHz.

The earliest applications of spread spectrum modulation were for military antijamming systems. In recent years, however, it has played a major role in wireless cell phone systems such as IS-95, CDMA, and 3G. The basic concept behind spread spectrum is code division multiple access, where each user is assigned a spreading different code. These codes may be entirely different or at a different phase position of the same code. All users occupy the same wideband spectra at the same time. At the receiver, the desired signal is recovered by multiplication of the appropriate $p_n(t)$ code. The desired signal bandwidth is collapsed, as mentioned before, while the product with other $p_n(t)$ signals results in yet another wideband signal. A narrowband filter recovers the desired signal. These operations are generalized in Figure 6.29. Now operation of spreading and dispreading does not change the power of the signal. It just rearranges the power in the frequency domain. By simple geometry, then, the areas under the signal blocks have the same power for all signals. But the collapsed signal has a bandwidth that is smaller than the spread signals

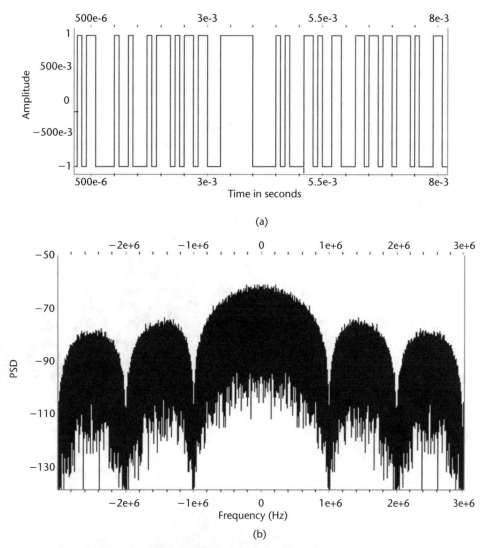

Figure 6.28   (a) Input data stream, and (b) its spectra. (c) The PSD of $p_n(t)$ spread signal showing CW interference at 500 kHz from center. (d) The collapsed spectra after $p_n(t)$ demodulation at the receiver. (e) The PSD of the collapsed data spectra after the $p_n(t)$ demodulation at the receiver. (f) The recovered data signal [compare with Figure 6.28(a)].

by the ratio of the original data rate to the spread rate B/W. So the output SNR_0 relative to the input SNR_i is simply

$$SNR_o/SNR_i = W/B$$

This ratio is called the processing gain (PG) of a spread spectrum signal. In the case detailed in Figure 6.28(a–f), this ratio is [1 MHz / 10 kHz] = 100 or 20 dB.

In the original military AJ operations, even a processing gain of 30 dB was not sufficient. This is due to the fact that the wrong user may be much closer to a receiver than the desired signal. This is the so-called near-far effect. Even after the dispreading operation, the unintended signal might still overpower the desired sig-

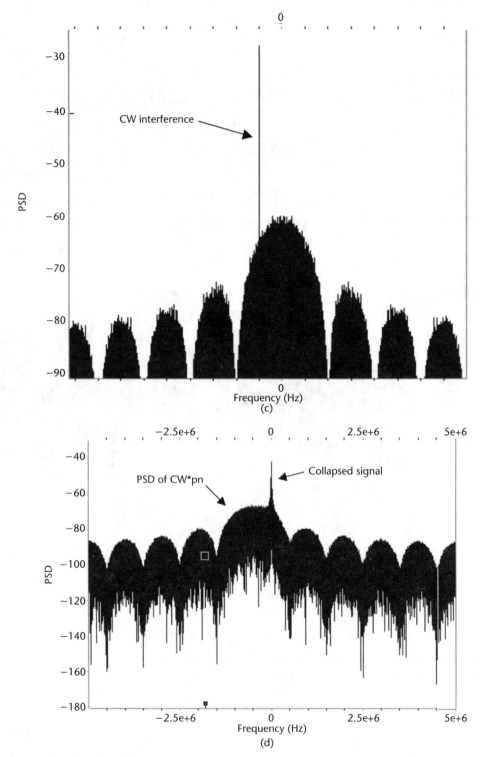

Figure 6.28 (continued).

nal. For this reason most of the tactical AJ systems such as Single Channel Air Ground Radio System (SINCGARS) use FH techniques. In the wireless world, it is

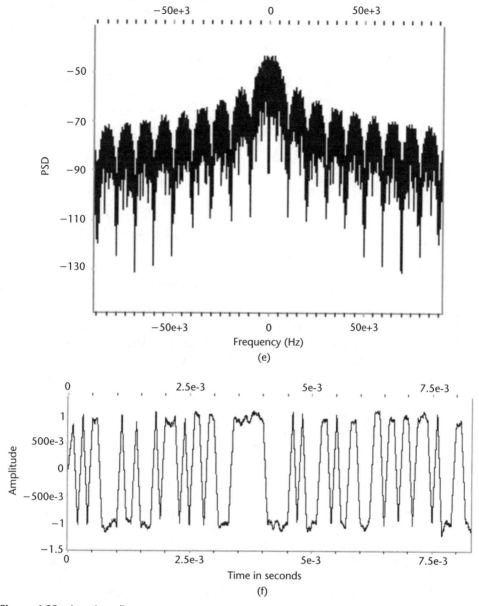

Figure 6.28   (continued).

possible (and mandatory) to control the user equipment power so that all signals arriving on top of each other at the base station have the same power.

One last thought: Suppose that the channel contains only AWGN. Then the spread spectrum operations produce a PG enhancing the system performance. In concept, we could increase the PG and correspondingly decrease the system BER. If only that were true. Recall in Chapter 5 on detection that for AWGN the output of the final matched filter is $2E_b/N_0$ without regard to the signal bandwidth. True, the spread spectrum operation as defined does give a PG, but the input SNR_i is lower by the same factor due to the spreading. We have opened up the receiver to a larger

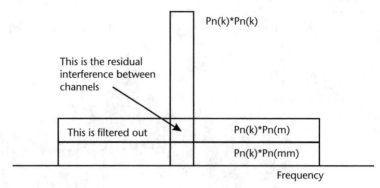

Figure 6.29   CDMA processing.

noise bandwidth *W*, not *B*. Thus, the final output SNR is the same with or without spread spectrum modulation.

6.13   Conclusion

In this chapter we presented a wide variety of modulation techniques. Note that they are all based on the fact that the only things we can change in the physical world are the amplitude, frequency, or phase of a sine wave. We also developed the basics of spread spectrum communications. These techniques were originally developed for military purposes, but now have found their way into many commercial systems, especially the wireless cell phone industry.

Selected Bibliography

Pahlavan, K., and A. H. Leveseque, *Wireless Information Networks,* New York: John Wiley & Sons, 1995.

Proakis, J. B., *Digital Communications,* New York: McGraw-Hill 1983.

Rappaport, T. S., *Wireless Communications, Principles and Practice,* Upper Saddle River, NJ: Prentice Hall, 2002.

Sklar, B., *Digital Communications, Fundamentals and Applications,* Upper Saddle River, NJ: Prentice Hall, 2001.

Steele, R., *Mobile Radio Communications,* London: Pentech Press Limited, 1992.

CHAPTER 7

Demodulation

Simulating the demodulator is considerably more difficult than simulating the transmitter. At the receiver, parameters such as frequency offset, bit timing, and symbol block time become significant. In this chapter we shall develop some basic demodulation procedures that are commonly and nearly universally employed. The last section is devoted to the concept of baseband simulation that eliminates the transmitted carrier frequency from consideration. This process can significantly increase the simulation speed.

7.1 In-Phase (I) and Quadrature (Q) Down Conversion Process

In Chapter 6 we showed that a transmitted signal can be represented by the form

$$s(t) = A(t)\sin(2\pi f_0 t + \theta(t))$$
$$= I(t)\cos(2\pi f_0 t) + Q(t)\sin(2 f_0 \pi t)$$

where f_0 is the transmitted carrier frequency, and the information is conveyed by the in-phase signal $I(t)$ and the quadrature signal $Q(t)$. The receiver tunes to f_0 filters the signal and usually produces the result to an intermediate frequency (IF). This is called a heterodyne receiver. There are several standard IF frequencies in use: commercial AM uses 455 kHz and commercial FM uses 10.7 MHz. Other well-known IF frequencies are 70 MHz and 21.4 MHz, and there are others.

The I/Q down conversion process is straightforward. The operation is shown in Figure 7.1. The output of this process is called I and Q, but they are not the information I and Q we started with. We label them $I'(t)$, $Q'(t)$, and the development below relates them to the original $I(t)$, $Q(t)$. The mathematical operations for $I'(t)$ are

$$I'(t) = \left[s(t)\cos(2\pi f_d t + \alpha) \right] * h(t)$$
$$= \left[I(t)\cos(2\pi f_0 t) + Q(t)\sin(2\pi f_0 t) \right]\left[\cos(2\pi f_d t + \alpha) \right] * h(t)$$
$$= \left[I(t)\cos(2\pi f_0 t) + Q(t)\sin(2\pi f_0 t) \right]\left[\cos(2\pi f_d t)\cos\alpha - \sin(2\pi f_d t)\sin\alpha \right] * h(t)$$

In the above, we have allowed for the possibility that the signal is not exactly centered in the IF band pass, $\Delta f = f_0 - f_d$, and there is an unknown phase difference α between the I Q oscillators and the phase of the received signal. The role of the filter $h(t)$ is to remove the mixing term with frequency $2f_d + \Delta f$ without disturbing the down converted or baseband signals I and Q. The results for $I'(t)$, $Q'(t)$ are

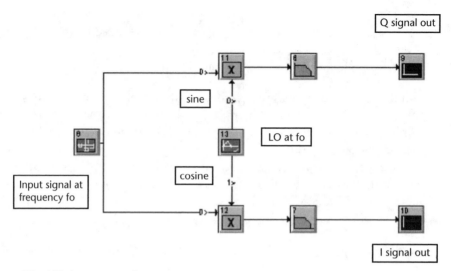

Figure 7.1   I/Q down conversion process.

$$I'(t) = I(t)\cos(2\pi\Delta ft + \alpha) + Q(t)\sin(2\pi\Delta ft + \alpha)$$
$$Q'(t) = I(t)\sin(2\pi\Delta ft + \alpha) - Q(t)\cos(2\pi\Delta ft + \alpha)$$

From here, $I'(t)$, $Q'(t)$ are processed in accordance with the type of modulation to recover $I(t)$ and $Q(t)$.

7.2 Low-Pass Filtering

The architecture of Figure 7.1 is nearly universal in modern day signal processing. As shown, it represents an analog implementation. However, the digital world has taken over and the operation is performed digitally in one of two ways. In the first method the signal is analog until after the low pass filters. At that time A/D circuits are inserted in the I and Q paths and convert the analog signals into digital I and Q signals. The subsequent demodulation operations are then performed digitally.

The second method makes use of the bandpass sampling concept introduced in Chapter 3. In this case the signal is still on a carrier but is in digital form.

The most common situation is an $F_s/4$ system where the carrier frequency is 1/4 the digital sample rate as described in Chapter 3. After the I/Q down conversion, we need a low pass filter to eliminate the sum frequency term at $2f_0$. This filter bandwidth needs only to be great enough to pass the signal without harm. There is another important consideration. Suppose the system sample rate is 100 Msps, and the required bandwidth is only 1 MHz. Then, by our discussion on Nyquist sampling, we are over sampled by factor of 50. One way to reduce the sample rate would be to simply decimate the filtered data stream by that amount. This is easy, but not smart. We are throwing away 49/50 or 98% of the computed filtered data. The solution is a decimating filter that only calculates the required data. This is an enormous savings in the simulation execution time.

7.2.1 CIC Filter

The cascaded integrator and comb (CIC) filter is a popular filtering technique that has been implemented in many forms. The idea is simple and is based on the structure shown in Figure 7.2. The structure is a series of digital integrators followed by a decimate-by-M operation, followed by an equally long series of equally long digital derivatives, and finally a compensating filter.

Relative to the output sample rate, the frequency response of the filter is

$$H(f) = \left[\sin\left(\pi f/f_s\right) / \sin\left(\pi f / M f_s\right) \right]^N$$

where N is the number of integrators/derivatives. $M = 2$ in Figure 7.2. The compensating filter is used to flatten out this response in the pass band.

7.2.2 Polyphase Filter

The polyphase structure is best presented by a direct example, specifically a decimate-by-2 FIR filter. The general output y to an input x processed by a filter h is the convolutional formula

$$y_k = \sum_{p=0}^{N-1} h_p x_{k+p}$$

Term by term, this expression, after deleting every other output, becomes

$$y_0 = h_0 x_0 + h_1 x_1 + h_1 x_2 + h_0 x_3$$
$$y_2 = h_0 x_2 + h_1 x_3 + h_1 x_4 + h_0 x_5$$
$$y_4 = h_0 x_4 + h_1 x_5 + h_1 x_6 + h_0 x_6$$
$$\ldots$$

For simplicity we used an $N = 4$ tap FIR filter (which does not effect the results), and we have omitted the odd values of y since we are decimating by 2. From the equation above, we see that the even numbered filter taps only see the even numbered data values. In the same way, the odd taps only see the odd data values. This suggests that we can arrange our processing as shown in Figure 7.3

This idea can be extended for higher order decimation. For $M = 4$ the demodulator would split the data in four paths, with each path being filtered by 1/4 of the coefficients. This idea can be extended for higher orders of decimation.

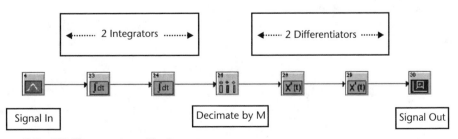

Figure 7.2 CIC filter structure, $N = 2$.

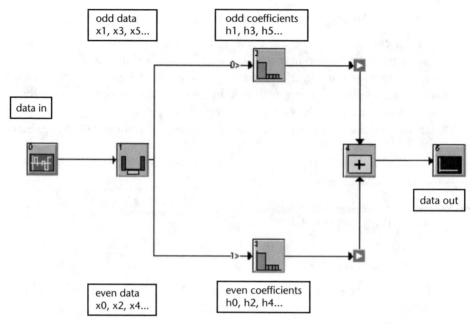

Figure 7.3   Polyphase decimate-by-2 filter.

7.2.3   Half Band Filter

A half band filter is defined through the relation

$$H(z) = c + z^{-1}f(z^2)$$

The constant c is usually taken to be $c = 0.5$. The first observation is that except for $n = 0$, all the even order coefficients are zero. If we are clever, we can perform the filtering without blindly multiplying by them. The second feature is that the transfer function has symmetry about the frequency $F_s/4$. Figures 7.4 and 7.5 show these features.

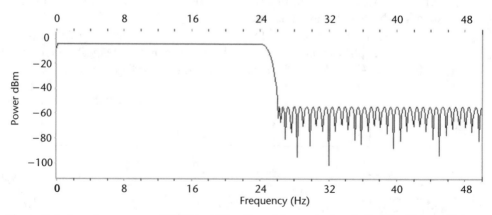

Figure 7.4   Impulse response of half band filter. Except for center tap, half of the filter coefficients are zero.

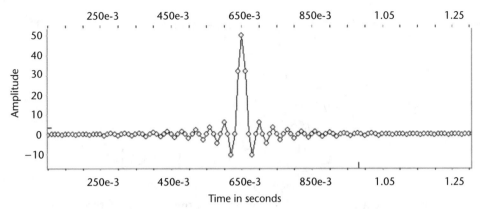

Figure 7.5   Frequency response of half band filter.

From Figure 7.5, we see that the filter transfer function is approximately a brick wall at $F_s/4 = 25$ Hz in this case. As the number of filter taps increase, the wall becomes more perfect. Using the polyphase concept just described and the fact that every other tap equals zero, we can implement an extremely efficient decimate-by-2 filter. If further sample rate reduction is required, we can just add another such filter (with the same coefficients) to the output of the first, and so on.

7.3   Offset Parameter Recovery

In order to demodulate the signal, the receiver must acquire three parameters from the received signal. First is the unknown frequency offset Δf, which can be caused by Doppler shifts due to motion, or the fact that the transmitter oscillator is not exactly equal to that of the receiver. Second is the unknown phase α relation between the transmitted and received signals. Even in a perfect world with absolute frequency control, the phase as seen by the receiver is shifted from that of the transmitter by a $= 2\pi f_0 \tau_d$, where $\tau_d = d/c$ is the transmission delay. Finally we need timing information. Where does an information bit or data block begin and end?

7.3.1   Frequency Offset

As pointed out in Chapter 1, in the I/Q down conversion process the down converting frequency is not necessarily the same as the center frequency at bandpass. This offset Δf must be eliminated or at least reduced to the point where it will not affect the demodulation.

Given an I/Q data stream, how can we calculate the associated frequency? By definition,

$$\Delta f = \frac{1}{2\pi} \frac{d\varphi}{dt} = \frac{1}{2\pi} \frac{d\left[\tan^{-1}(Q/I)\right]}{dt}$$

But the inverse tangent algorithm is time consuming, and it is generally avoided if possible. To this end we further develop the definition

$$\Delta f = \frac{1}{2\pi}\frac{d\varphi}{dt} = \frac{1}{2\pi}\frac{d\big[\tan^{-1}(Q/I)\big]}{dt}$$

$$= \frac{1}{d\pi}\big[Q\,dI/dt - I\,d\,Q/dt\big]\big/\big[I^2 + Q^2\big]$$

$$dI/dt = (I_n - I_{n-1})/\Delta t$$

$$dQ/dt = (Q_n - Q_{n-1})/\Delta t$$

$$\Delta f = \frac{1}{2\pi}\big[Q_n I_{n-1} - I_n Q_{n-1}\big]\big/\big[I_n^2 + Q_n^2\big]\Delta t$$

This is a neat result. The final estimate is the average of the last equation over all data points in a block of N. In some cases, such as MPSK, the envelope $I_n^2 + Q_n^2$ is constant and can be eliminated from the calculation.

The basic problem with the above methods is that the differentiation process is *noisy*. The general expression for the signal including modulation $m(t)$ is

$$\bar f = \overline{1/2\big[d\varphi/dt\big]} = \overline{1/2\,d\big[2\pi f_0 t + m(t)\big]/dt}$$

$$= f_0 + \overline{1/2\,dm(t)/dt}$$

Thus, the frequency estimate has a noise term, which is the average of the modulating frequency. Consider the simple case where the modulating signal is MPSK. Then every time the calculation straddles a bit transition, there can be a jump discontinuity in the output, as shown in Figure 7.6.

Two other methods are based on the frequency domain PSD of the signal. The first method is just to pick the maximum point of the PSD and choose that frequency as a result.

Figure 7.7 shows the difficulty of using this process. The FFT is taken over a finite set of signal data. Thus the PSD can be ragged as shown. It is possible to smooth the result by widowing or by averaging shorter FFT segments of the data.

The second method is to integrate the PSD and choose the point where that integral is at 1/2 of the maximum.

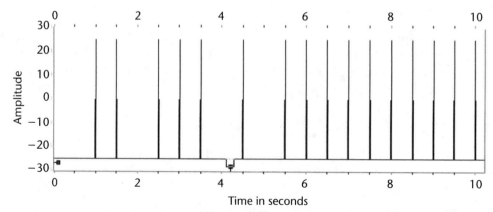

Figure 7.6 Frequency estimation of a BPSK. Note the spike where the data modulation switches phase.

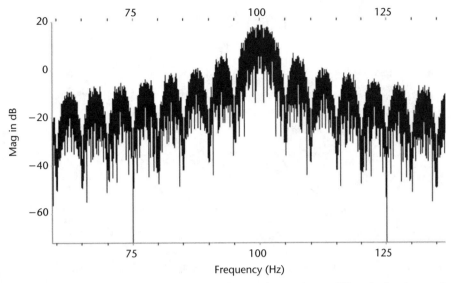

Figure 7.7   Frequency estimation from the signal PSD. The maximum PSD point is taken as the offset.

Figure 7.8 is the integral of the PSD of Figure 7.7. The integrator itself acts like low pass filter. The rise of the curve is also somewhat ragged, but taking the 50% point is usually more accurate.

A final method is the slope of the signal phase. In general, the output of the phase detector can be written as

$$\varphi(t) = 2\pi\Delta f t + m(t)$$

where $m(t)$ is due to the modulation. But $\overline{m(t)} = 0$, so if we fit this phase to the best straight line, the slope will provide the frequency estimate as shown in Figure 7.9. In the MPSK case, we still get the jump discontinuities, but their effect is much less in this calculation. Note: When using the arctan function to get the phase, you must use the option that unwinds the phase. Normally the arctan will produce an answer

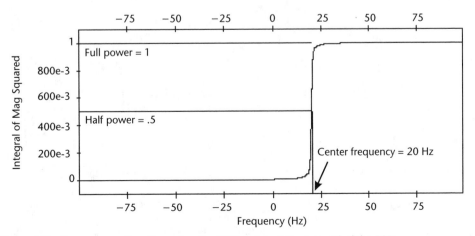

Figure 7.8   Frequency estimation using the 50% point of the integral of the PSD.

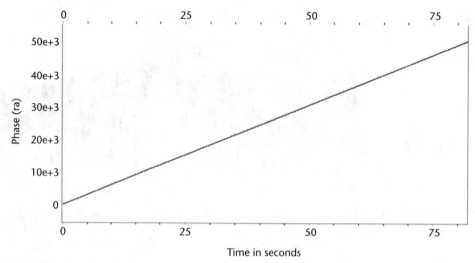

Figure 7.9 I/Q phase slope due to frequency offset. Fit the best straight line to get the offset estimate.

between $\pm\pi$, so without the unwinding operation the phase plot would look as shown in Figure 7.10. The unwinding algorithm anticipates when the phase crosses the $\pm\pi$ boundaries, and simply tacks on the appropriate value so the track is a straight line.

The above algorithms are called open loop estimates. The results are obtained by operating on a block of data in total. Another method, known as closed loop, is to implement a tracking or automatic frequency control (AFC) loop as shown in Figure 7.11. Figure 7.12 shows the error function as the loop locks.

7.3.2 Data Timing

Now that we have removed the frequency offset, the next step is to obtain bit sync, or where the data transitions occur. Figure 7.13 shows a block diagram of such a tracking circuit.

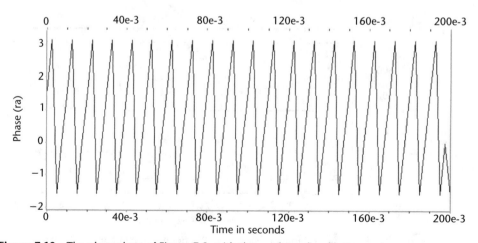

Figure 7.10 The phase slope of Figure 7.9, with the result confined between $\pm\pi$.

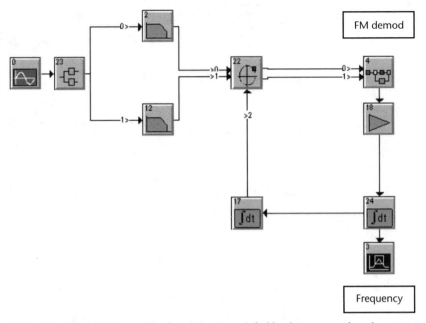

FM demod

Frequency

Figure 7.11 Wideband AFC loop. The frequency error is fed back to a complex phase rotator to provide correction.

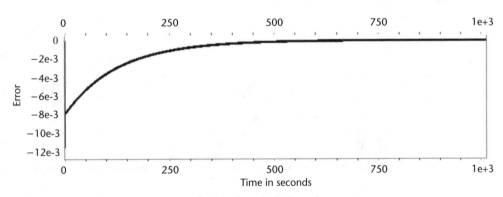

Figure 7.12 The frequency error as the AFC loop locks in and goes to zero.

This is called an absolute value type loop. An error signal is obtained from the input data from

$$e(\tau) = \left| \int_0^{T/2} s(t - \tau)dt \right| - \left| \int_{T/2}^{T} s(t - \tau)dt \right|$$

where τ is the unknown offset with respect to $t = 0$. Clearly, if $\tau = 0$, $e(0) = 0$ as well. Figure 7.14 shows the error function as the loop locks.

Figure 7.15 shows the initial timing mismatch between the sampler and the matched filter output, and Figure 7.16 is the same plot only when the loop is locked.

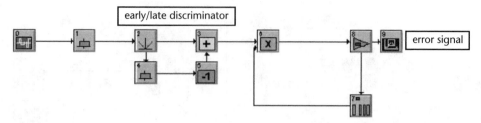

Figure 7.13   Absolute value early-late gate data sync block diagram.

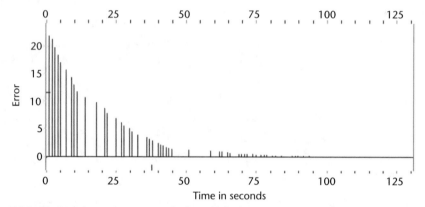

Figure 7.14   Typical data sync error as the loop locks in.

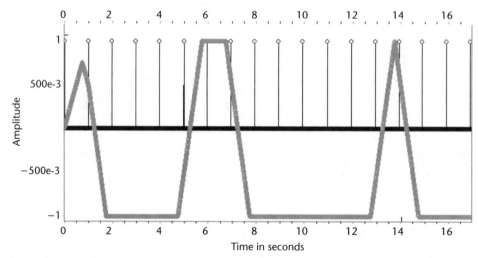

Figure 7.15   Initial timing offset of the local clock with respect to the received data transitions.

7.4   Phase/Frequency Recovery

In Section 7.3.1 we showed methods for obtaining the frequency of a modulated signal. These algorithms all fall into the category of wideband AFC techniques. In this section we describe phase frequency acquisition techniques based on a phase error approach.

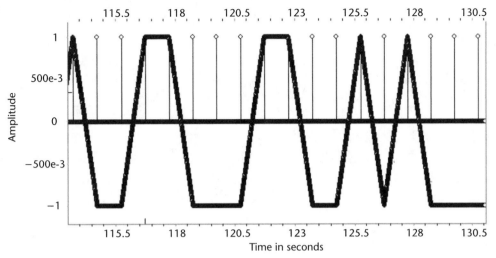

Figure 7.16   Final timing alignments when loop is locked. The samples occur at the top of the data.

7.4.1   BPSK

For BPSK the commonly used method is the Costas loop. The idea is quite simple. Consider a BPSK signal in the form. The basic block diagram is shown in Figure 7.17.

Figure 7.18 shows the loop locking onto the proper input frequency offset.

Figure 7.19 shows the I channel signal as the loop locks, and Figure 7.20 shows the a channel being rotated out.

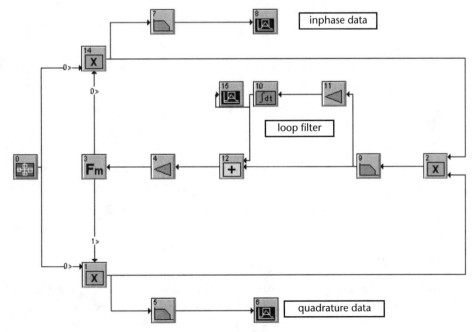

Figure 7.17   Costas loop block diagram.

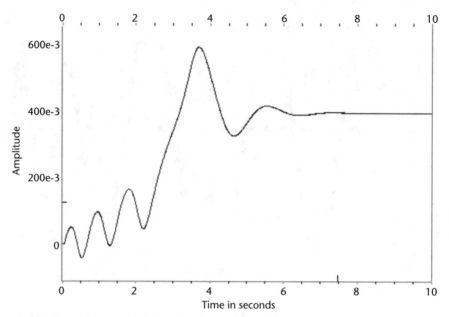

Figure 7.18   Costas frequency offset estimation as loop locks.

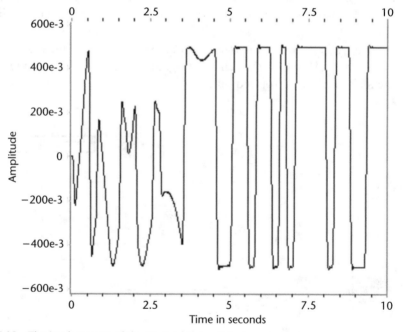

Figure 7.19   The in-phase arm of the Costas loop as it locks.

7.4.2   QPSK

For a QPSK signal, the trick used for the Costas loop (that I*Q is independent of the modulation) no longer works since there are four phases. One possibility is to extend the idea of the squaring loop to a fourth power loop. This will produce a carrier at four times the nominal signal frequency. While this works analytically, in

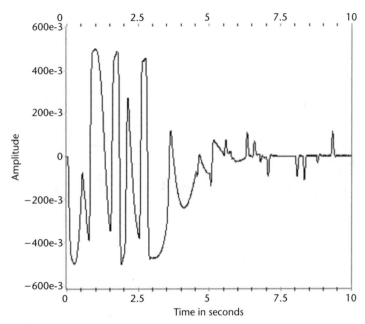

Figure 7.20   The quadrature arm of the loop as it locks.

practice it forces the system sample rate up by that same factor so as to not alias the recovered tone. This is not a good thing.

A very effective variant on the Costas idea is to compute an error signal based on the relation

$$e(t) = I(t)\big|Q(t)\big| - Q(t)\big|I(t)\big|$$

Figure 7.21 shows the error as a function of the phase offset. Note that there are now four stable tracking points as opposed to the two points in the Costas loop. Again, either an absolute reference must be established, or the phases must be differentially encoded at the transmitter.

Figure 7.22 is the general block diagram of the QPSK loop under discussion. Figure 7.23 shows the loop locking onto the input frequency offset.

7.4.3   Decision Feedback

The QPSK algorithm is a special case of the more general concept of decision feedback. This method can be used on higher order MPSK or even QAM modulation. The basic idea is to compute an error that is the difference between the actual received I/Q signal, and the nearest neighbor constellation point associated with the modulation format.

7.5   Conclusion

In this chapter we developed the techniques and algorithms required to demodulate a signal and recover the information. The basic issue is to recover the carrier phase,

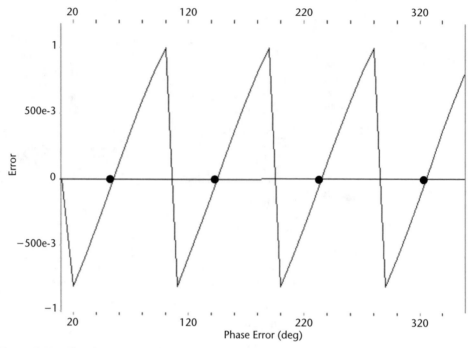

Figure 7.21   The phase error of the QPSK signal. Note that there are four stable tracking points.

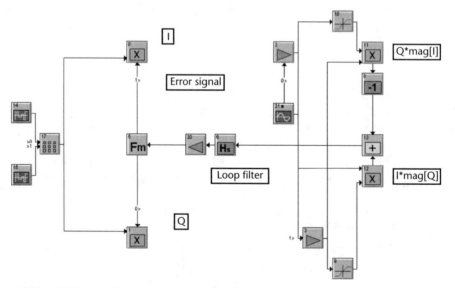

Figure 7.22   QPSK phase/frequency recovery algorithm, block diagram.

unknown phase offset, and symbol timing. The starting point for all of these algorithms is the down converted I and Q signals.

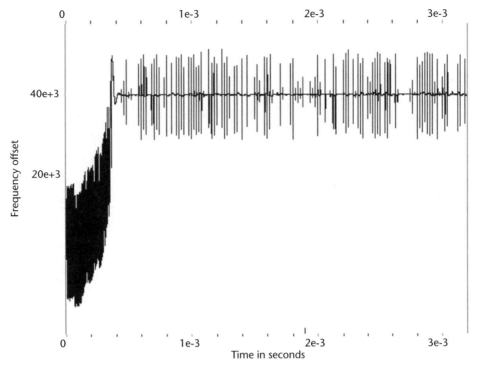

Figure 7.23   The frequency offset estimate of the QPSK loop as it locks.

Selected Bibliography

Pahlavan, K., and A. H. Leveseque, *Wireless Information Networks*, New York: John Wiley & Sons, 1995.

Proakis, J. B., *Digital Communications*, New York: McGraw-Hill 1983.

Rappaport, T. S., *Wireless Communications, Principles and Practice*, Upper Saddle River, NJ: Prentice Hall, 2002.

Sklar, B., *Digital Communications, Fundamentals and Applications*, Upper Saddle River, NJ: Prentice Hall, 2001.

Steele, R., *Mobile Radio Communications*, London: Pentech Press Limited, 1992.

Vaidyanathan, P. P., *Multirate Systems and Filter Banks*, Englewood Cliffs, NJ: Prentice Hall, Inc., 1993.

CHAPTER 8

Baseband Pulse Shaping

In Chapter 3 we noted that there is no such thing as a perfectly band limited signal. This fact has serious consequences in many places. For example, frequency is an economic commodity. The United States government regulates who can broadcast what signals in what bands. During the wireless revolution, for example, a bidding process was established to accommodate the various services providers. The highest bidder received the license to transmit their signal. In order to maximize the return on investment, the winner wants to serve as many customers as possible. In general, the allocated bandwidth is segmented into channels with a given group of subscribers assigned to a particular slot. One of the main problems is to avoid one channel from interfering with the next. This phenomenon is called adjacent channel interference. Here is the problem: Suppose that the amount of power spectral density generated by a signal in slot A is, say, 80 dB below the slot A carrier (dBc) as it spills into the next frequency slot B. Now suppose you are user B. You are talking to a transmitter source quite far away while the user on slot A is much closer. Then it can (and does) happen that the weak amount of spill from frequency slot A into slot B is greater than the direct signal power in frequency slot B. Slot B is blocked by slot A. This is called the near-far effect. Another system that suffers in the same manner is air traffic control signals around an airport where one plane may be just in range of say 100 miles while a second plane may be just touching down on the runway. A third system is Army tactical radios where one transmitter may be over the hill talking to a command post that has several different transmitters actively talking to others.

In this chapter, we develop the commonly employed band limiting process for digital transmission known as baseband pulse shaping.

8.1   Baseband Pulse Analyses

Consider the simplest representation of a digital waveform known as non return to zero (NRZ) having the form

$$s(t) = \sum_k c_k h_{NRZ}(t - kT)$$

where $c_k = \pm 1$, T is the symbol time = 1/symbol rate, and the NRZ pulse is given by

$$h_{NRZ}(t) = 1 \quad 0 \le t \le 1$$
$$= 0 \quad \text{otherwise}$$

Figure 8.1 shows a typical result for $s(t)$.

Now let us examine the power spectral density (PSD) of $s(t)$:

$$R_{NRZ}(f) = \lim_{n \to \infty} \left\{ avg \left[\left| R_n(f) R_n^*(f) \right|^2 / n \right] \right\}$$

$$R_n(f) = \int_{-nT}^{nT} s(t) e^{-2\pi ift} \, dt$$

$$= \int_{-nT}^{nT} \sum_k c_k h_{NRZ}(t - kT) e^{-2\pi ift} \, dt$$

$$= \sum_k c_k \int_{-nT}^{nT} h_{NRZ}(t - kT) e^{-2\pi ift} \, dt$$

$$= \sum_k c_k e^{-2\pi ikfT} \int_0^T e^{-2\pi ift} \, dt$$

$$= \left[\left(e^{-2\pi ifT} - 1 \right) / 2\pi if \right] \sum_k e_k e^{-2\pi ikfT}$$

In the above, the avg operation is over all possible combinations of the data bits c_k. Under the assumption that these bits are statistically independent, the final result for the PSD is

$$R_{NRZ}(f) = \left[\sin(\pi fT) / \pi fT \right]^2$$

Figure 8.2 is a plot of this function for $T = 1$.

Note that this spectrum dies off very slowly beyond $f = 1/T$. A good system number to remember is the relative power of the first spectral lobe at $f = 3/2T$ with respect to the PSD evaluated at $f = 0$. The result is −13.8 dB. At the second lobe at $f = 5/2T$, the power drops to only −17.9 dB. To put this into some perspective, consider two transmitters whose frequencies are $5/2T$ apart. Now assume that the RF propagation power dissipates as $1/\text{distance}^2$. Then for two equal power signals, at a relative distance up to a factor of 7.88, the second side lobe power of the stronger signal, slot A, will equal or exceed the direct power in slot B. This is shown in Figure 8.3.

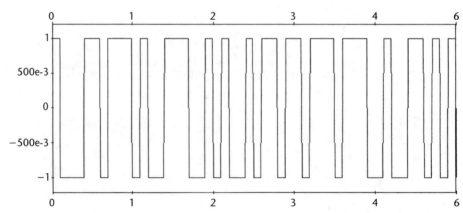

Figure 8.1 Basic NRZ waveform.

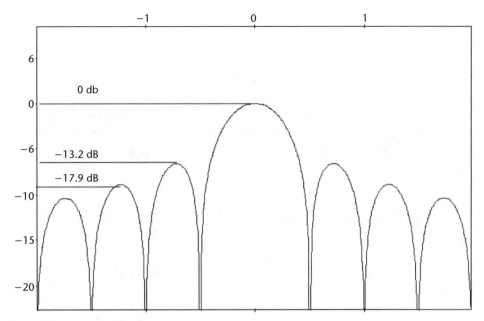

Figure 8.2 PSD of NRZ pulse.

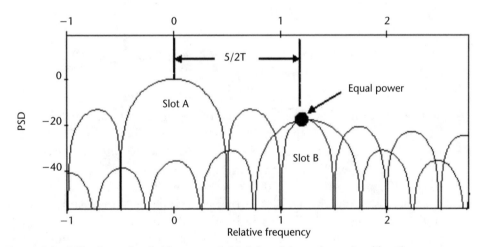

Figure 8.3 PSD of two signals. The spectral side lobe of the stronger signal has the same power as the main lobe of the second signal.

So we see that the NRZ pulse has an energy spectra that *splashes* all over the place. What we desire is a PSD that is absolutely confined to some bandwidth B. What happens if we reverse the role of the NRZ pulse and its corresponding sinc function spectra? Then the pulse shape $p(t)$ is given by

$$h_{\sin c}(t) = \sin(\pi t/T)/(\pi t/T)$$

Then the spectral occupancy is

$$R_{SINC}(f) = 1 \quad |f| = 1/2T$$
$$= 0 \quad \text{otherwise}$$

This is great except, as pointed out in Chapter 3, the sinc function is infinite in extent and does not exist in the real world. This means that to realize this function it must be truncated at some finite value $\tilde{T}$. Now the sinc function falls off very slowly as $1/t$, so it takes a large $\tilde{T}$ to avoid serious degradation from the optimum performance. This is the dual statement to the slow frequency roll-off of the PSD of the NRZ pulse.

8.2 The Raised Cosine Pulse

There are many possible choices for $p(t)$ which serve as a good compromise between the two examples just considered. The most popular is the raised cosine pulse given by

$$h_{RC}(t) = \sin c(t/T) \cos(\pi \alpha t/T)/(1 - 4\alpha^2 t^2/T^2)$$

where $0 \leq \alpha \leq 1$ is the roll of or excess bandwidth factor. The PSD of this pulse is

$$R_{RC}(f) = 1 \qquad\qquad\qquad 0 \leq f \leq (1-\alpha)/2T$$
$$= 5(1 - \sin\{\pi T(f - 1/2T)a\}) \qquad (1-\alpha)/2T \leq f \leq (1+\alpha)/2T$$
$$= 0 \qquad\qquad\qquad\qquad \text{otherwise}$$

When $\alpha = 0$, this function reverts back to the sinc function, which has a maximum bandwidth of $1/2T$, which is the smallest possible. When $\alpha = 1$, the maximum bandwidth is $1/T$, which is double the minimum value, giving rise to the term excess bandwidth. Figure 8.4(a) shows this pulse shape for a typical roll-off factor of $\alpha = 0.3$, and Figure 8.4(b) shows the corresponding PSD.

The observant reader will note that $h_{RC}(t)$ is just as unrealizable as the sinc function. So what have we gained? The answer is that $p_{RC}(t)$ falls off in time as $1/t^3$, which is much faster than the $1/t$ roll-off for the sinc function. Thus, while still necessary, the truncation can be accomplished for a much smaller time duration of the wave form. Figure 8.5 shows the overlaid spectra for three cases with a data rate of 10 Hz. The first is the sinc arising from the NRZ used as a reference. The second is the spectra from the sinc(t) function which has been truncated to $\pm10T$. The third is the spectra of the raised cosine pulse truncated to the same degree. Now, in the nonexistent perfect world, the spectra of the sinc pulse would be constant to .5 Hz and zero beyond (a brick wall filter). In the same manner, the RC spectra would smoothly role off. What Figure 8.5 shows is the sinc pulse spectra starting to drop at 5 Hz, but then splattering out. The RC spectrum, at the beginning, emulates the theoretical curve, but it also splatters out in frequency. Note that the RC splatter is much lower that the sinc splatter. This is because for the truncation time $10T$, the RC impulse response is much lower in amplitude (less residual loss) than the sinc pulse. Figure 8.6 shows an overlay of these two time functions.

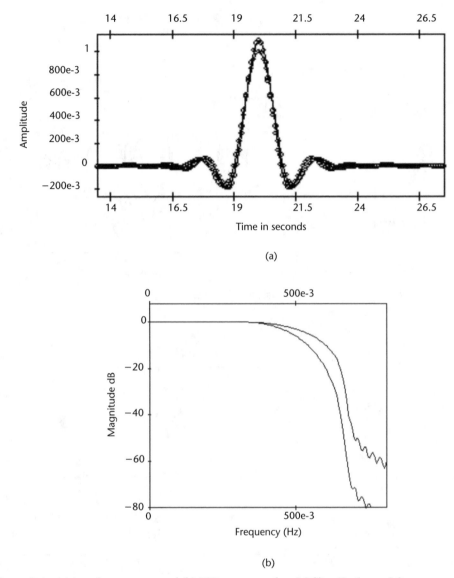

Figure 8.4   (a) Impulse response and (b) PSD response of an RC filter $T = 1$, $\alpha = 0.3$.

How do you simulate a signal with a raised cosine pulse, or any other pulse shape for that matter? The answer is obtained by observing the following relation for arbitrary pulse shape $h(t)$:

$$s(t) = \sum_k c_k h(t - kT)$$
$$= \sum_k c_k \delta(t - kT) * h(t)$$

What this says is that you generate a series of impulses separated by time T, of proper weight c_k, and run them into a filter whose impulse response is $h(t)$. Figure 8.7 shows this operation, and Figure 8.8 shows the resulting pulse train required to impulse the baseband pulse filter.

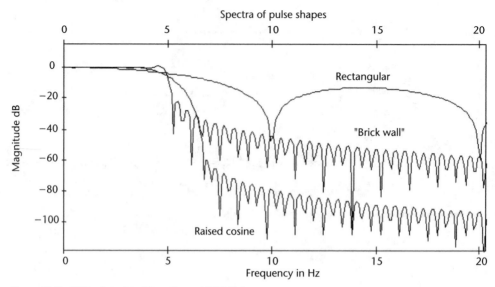

Figure 8.5   PSD of sinc, brick wall, and RRC filters.

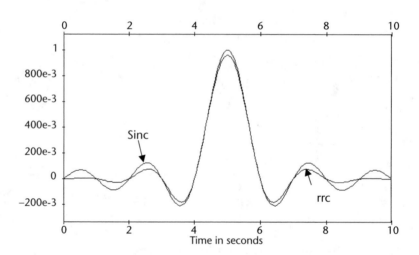

Figure 8.6   Overlay of the RRC and sinc pulse for $t = 1$ and $\alpha = 0.3$.

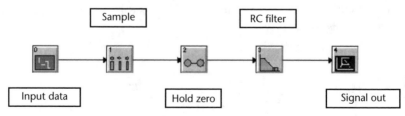

Figure 8.7   Simulation of a pulse shaped data stream using impulse functions.

What if impulses are hard to come by, and what happens when you just run the NRZ waveform into the pulse filter? In that case the output spectrum $H(f)$ will be

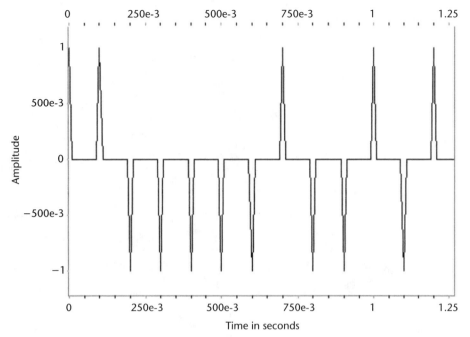

Figure 8.8 Impulse string used to excite the pulse shaping filter.

the product of the filter spectra $R(f)$ and the sinc function, which is the transform of the NRZ pulse. It follows that the ISI properties for the RRC filter are destroyed. There is one trick available. Insert the inverse of the sinc filter in the processing chain to recover the desired spectra. In the Laplace domain, the sinc filter has the transfer function

$$H(s) = \int_0^T [1] e^{-st}\, dt = \left(1 - e^{-sT}\right)\big/ s$$

so the inverse function has the transfer function

$$H^{-1}(s) = s\big/\left(1 - e^{-sT}\right)$$

Figure 8.9 shows how to simulate this function. The "s" is a derivative, and the denominator is a feedback with a T-second delay. This model only works in the simulation if the sample rate is an integer multiple of (commensurate with) the data rate T.

8.3 Intersymbol Interference and Eye Diagrams

Let us return to the basic expression

$$s(t) = \sum_k c_k b(t - kT)$$

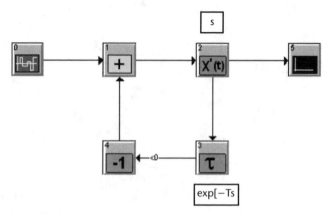

Figure 8.9   Simulation of the inverse sinc function.

What we would like to have happen is for $s(kT) = c_k$. From the above expression, this can only happen if

$$h(kT) = 0 \quad k \neq 0$$

For an arbitrary pulse shape, $h(t)$, this might not be the case, and the actual result would be

$$s(0) = c_0 + \sum_{k \neq 0} c_k h(kT)$$

What is happening is that the attempt to recover the signal at $t = 0$ is corrupted by the other data values via the impulse response of $h(t)$. This phenomenon is called intersymbol interference. It is a major problem in many communication systems.

To continue the discussion we introduce the concept of an eye diagram. Suppose we generate a signal using the raised cosine pulse and print it on a long piece of transparent paper. On this paper we make marks at time intervals of T seconds. We fold the paper up like an accordion using these markers so that each segment lies on the other and is visible. Figure 8.10 shows such a plot but with the markers set at $2T = 0.2$ sec for a better visual presentation.

The shape of the resulting figure is the basis for the name. Another way of generating the eye diagram is with an oscilloscope. A T-second clock generates both the signal and the sweep trigger. By setting the scope persistence up, the same picture will emerge.

The most striking feature of the eye diagram curve of Figure 8.10 is the sharp convergence of the eye to two points (± 0.1) at multiples of $T = 0.1$ sec. This is a manifestation of the fact that the raised cosine pulse, like the sinc pulse, is equal to 0 at multiples of $t = kT$, $k \neq 0$. Thus, at the proper sampling instant, only one pulse contributes to the output, and the rest disappear from the result, producing a condition with no ISI.

To see how ISI can be introduced in a communication system, consider Figure 8.11, which shows a basic communication link. The transmitted signal is $s(t)$ as before. At the receiver there is some form of receive filter, usually one that is matched to the transmit filter. The resulting received signal $r(t)$ is

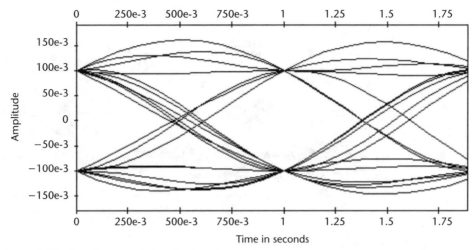

Figure 8.10   Eye diagram of RC pulse signal.

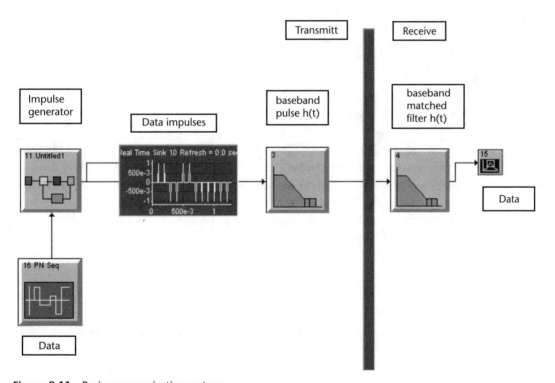

Figure 8.11   Basic communication system.

$$r(t) = s(t) * h(t)$$
$$= \sum_k c_k \int h(\tau - kT)h(t - \tau)d\tau$$
$$= \sum_k c_k g(t - kT)$$

Figure 8.12 shows the corresponding eye diagram for $r(t)$, where the basic pulse is $g(t) = h_{RC}(t) * h_{RC}(t)$.

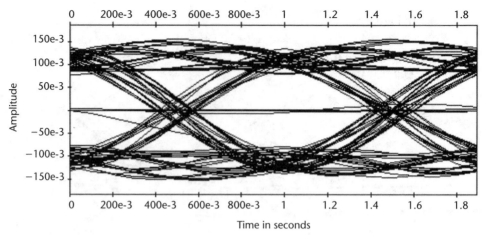

Figure 8.12 Eye diagram of $g(t) = h_{RC}(t)*h_{RC}(t)$.

What Figure 8.12 shows that the result of convolving the RC filter with itself introduces ISI, even though the individual components do not. The ISI is manifested in the eye starting to close at $T = 0.1$, and there are no longer two simple sharp points at $T = 0.1$ sec.

8.4 The Root Raised Cosine Filter

How do we fix the dilemma just derived for the RC filter that produces ISI when used in the receiver, as a matched filter to the one in the transmitter? The solution adopted by many wireless standards such as IS-136 and others is to split the RC filter into two haves, and put one half in the transmitter, and the other half in the receiver. The resulting filter is the raised root cosine (RRC) filter defined in the frequency domain as follows:

$$H_{RRC}^2(f) = H_{RC}(f)$$
$$H_{RRC}(f) = \sqrt{H_{RC}(f)}$$

This gives rise to the name RRC. A corollary to this development is

$$h_{RRC}(t) * h_{RRC}(t) = h_{RC}(t)$$

The impulse response of the RRC filter is somewhat complicated as given by

$$h_{RRC}(t) = k\left[(4\alpha/T)\cos(\Delta_+ t) + \sin(\Delta_- t)/t\right]/\left[(4\pi\alpha t/T)^2 - 1\right]$$
$$k = \sqrt{T}/\pi \qquad \Delta_\pm = (1 \pm \alpha)\pi/T$$

As before, this function is not realizable either and is subject to the same issues regarding truncating to some finite time that is sufficient.

In Figure 8.13, we show the eye diagram of a signal generated by the RRC filter. Note that there is considerable ISI at this point. But we do not care because this is not the signal that is used to recover the information.

The transmit and receive signals for this case are

$$s(t) = \sum_k c_k h_{RRC}(t - kT)$$

$$r(t) = s(t) * h_{RRC}(t)$$

$$r(t) = \sum_k c_k h_{RRC}(t - kT) * h_{RRC}(t)$$

$$r(t) = \sum_k c_k h_{RC}(t - kT)$$

Figure 8.14 shows the result of the convolution of the RRC filter with itself. Compare this with Figure 8.10.

Table 8.1 summarizes the ISI properties of the various pulse shaping filters considered here.

8.5 Conclusion

In this chapter we introduced the concept of baseband filtering. The object of these filters is to confine the occupied bandwidth of a signal so it does not splatter into adjacent bands. The RC filter was introduced as a commonly used filter. An analysis of optimum matched filter processing with the RC filter leads to the concept of ISI and the eye diagram. The RRC filter was derived as half of an RC filter placed in both the transmit and the receive side of a communication link. The RRC combination accomplishes two basic goals in communication system design. First, they act as a matched filter pair for optimum SNR detection. Second, the filter combination does not produce ISI.

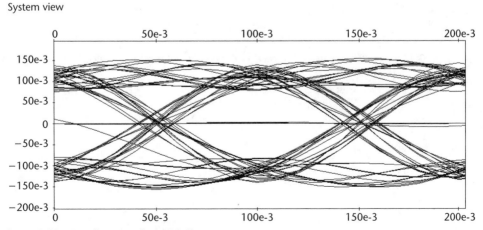

Figure 8.13 Eye diagram of an RRC filter.

System view

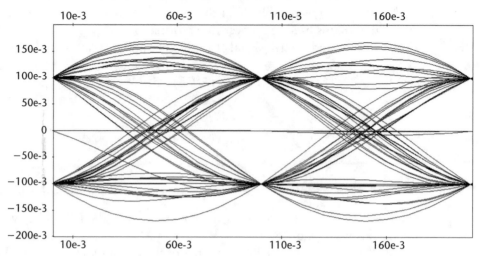

Figure 8.14   Eye diagram of an RRC filter convolved with itself.

Table 8.1   ISI properties of Baseband Filters

Pulse Shape	ISI at Transmitter	ISI After Matched Filter
Rectangular	No	No
Sinc	No	No
Raised cosine	No	Yes
Root raised cosine	Yes	No

Selected Bibliography

Pahlavan, K., and A. H. Leveseque, *Wireless Information Networks*, New York: John Wiley & Sons, 1995.

Proakis, J. B., *Digital Communications*, New York: McGraw-Hill 1983.

Rappaport, T. S., *Wireless Communications, Principles and Practice*, Upper Saddle River, NJ: Prentice Hall, 2002.

Sklar, B., *Digital Communications, Fundamentals and Applications*, Upper Saddle River, NJ: Prentice Hall, 2001.

Steele, R., *Mobile Radio Communications*, London: Pentech Press Limited, 1992.

CHAPTER 9

Bit Error Rate Calculations

In digital communications the near universal figure of merit for a system is the BER. As a function of the signal to noise or interference, what is the probability the data bit sent was received in error (i.e., $0 \rightarrow 1$, or $1 \rightarrow 0$)? A related measure that is sometimes used is the message error rate (MER). For example, the sent message STOP is not decoded correctly, but decoded to another possible message like GO. In this chapter we will develop the concept of a bit error and how to set up accurate BER measurements that run in the least time possible.

9.1 The BER Measurement Process

Figure 9.1 shows a simple overall statement of the BER process.

A data source $X_1(k)$ at R bps is encoded, modulated, and transmitted. The medium between the transmitter and receiver is known as the *channel*. The channel has two effects on the signal. First, it can simply add noise and other interference, and second, the signal can bounce off of objects and create a fade. The receiver demodulates the signal, establishes the proper timing for detection, and finally puts out a second data stream $X_2(k)$ of rate R, which the receiver believes to be the actual message. The BER is the bit-by-bit comparison of this data stream with the input stream. The measurement algorithm is the XOR logical operation

$$Y(k) = X_1(k) \otimes X_2(k)$$

This is the digital equivalent of a multiplication as shown in Table 9.1.

For a run of N bits, the BER is computed as

$$BER' = \sum_{k=0}^{N-1} Y(k)/N$$

So the first decision to be made is how big N must be. The larger N, the longer the simulation must run. To answer this, we note that the BER expression above is a statistical quantity. Due to the randomness of the noise and other factors, 10 runs of N trials will lead to 10 different results. Which one is the absolutely correct answer? Actually, none of them. That is the meaning of the accent mark on BER'; it is an estimate of the true BER. The operation on each trial produces a 1 with probability $p =$ true BER, and a 0 with probability $1 - p$. This type of statistical system is called the binomial distribution. From probability theory, the mean (expected value), μ, and standard deviation, ó, of the average on N such variables (i.e., BER') is

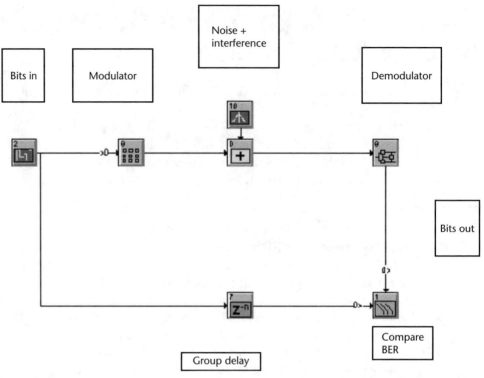

Figure 9.1 Basic communication system block diagram with BER measurement included.

Table 9.1 The XOR Operation

$X_1(k)$	$X_2(k)$	$Y(k)$
1	1	0
1	0	1
0	1	1
0	0	0

$$Ex(BER') = \mu = p = BER$$
$$std(BER') = \sigma = p(1-p)/N$$

Thus, the value measured value BER' becomes the true BER in the limit of large N, and the standard deviation of this estimate goes to 0. In accordance with the law of large numbers, the probability distribution of BER' can be written in the form

$$p_{BER'}(x) = e^{-(x-\mu)^2/2\sigma^2}/\sqrt{2\pi\sigma^2}$$

From this form, one can use standard hypothesis testing choose N such that the probability of the measured result is within, say, 0.001 of the true value is a prescribed amount, say, 99%. Figure 9.2 illustrates this concept. The system was set for

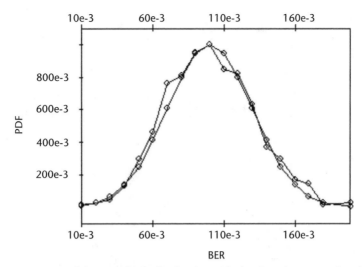

Figure 9.2 Comparison of the actual BER distribution with the Gaussian approximation formula.

$p = 0.1$ and $N = 100$. The histogram overlays the actual histogram of the BER result variation for 1,000 trials, overlaid with the corresponding Gaussian formula.

Like many nice results, this has some problems. In particular, we must know p to begin with, which we usually do not. One could do a relative short run to get an estimate. In the end result, there is the compromise between run time and accuracy. The author usually starts with $N = 10$, and modifies it if more accuracy is needed.

9.2 System Group Delay

Many of the operations commonly used in communication systems have a processing or group delay. The group delay has several causes in a simulation. One source is the filter group delay as described in Chapter 6. If there are several filters in tandem, the group delay adds. Block forward error correcting (FEC) coding serves as a second example. The FEC encoder takes in a group of bits k, and calculates an output code word with N bits. Suppose that the input block covers T seconds. Then for the first T second, the encoder is gathering the data. At T seconds, the encoder stops and calculates the N output bits. These bits are presented to the system from $T < t < 2T$ seconds. At the decoder, the procedure repeats itself in reverse. The FEC decoder must know where a block of bits begins and ends. This is part of the synchronization process. However, once found, the decoder grabs T seconds of data and halts the simulation again while it decodes the k recovered bits. The overall delay is now $2T$. Others include bit-to-symbol converters, and Gray encoders.

How, then, do we calculate this group delay? One choice would be to calculate the contribution of each element in the system, and add them up to get the total. In a large system this procedure would not be remotely feasible even if you could find and calculate which element contributed what delay. A simple and effective way is to simply cross-correlate the input data stream with the output. The time location of the maximum of this cross-correlation is the required offset. Figure 9.3 (a–c) shows this concept. The system group delay is 5 data bits. The cross-correlation peak is at

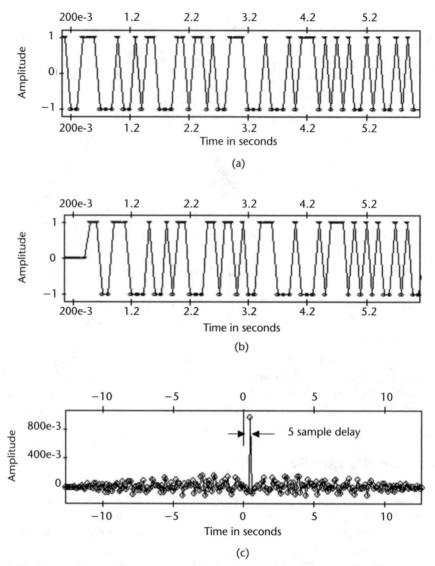

Figure 9.3 (a) Input data; (b) output data [compared with (a), the data isdelayed by five samples]; and (c) the cross-correlation between the signal in and out. The peak maximum is at five samples from zero.

five samples. Note, it does not matter whether you do a*b or b*a. The difference between these two results is just the algebraic sign on the result. But, from the system, we know that the input must be delayed to match the output, so only the absolute value of the peak location is required.

Referring back to Figure 9.1, the group delay is first set to 0. The input and output data streams are cross-correlated (with all noise and interference removed) and the location of the maximum is determined. The BER that is reported at this point should be about 0.5. Now enter this offset and repeat the BER and cross-correlation. The peak of the cross-correlation should now be at 0 offset, and the BER should be 0. It is wise in these measurements to delay the BER measurements by at least the group delay to allow for initial system transients and to give time for other processes to settle down to a steady state.

In this example it was assumed that the system delay was exactly 5 bits in length. This means that the input data is on time makers 0, T, $2T$, ..., and the output data is on time makers $4T$, $5T$, ..., and so on. But this is not always the case. In fact, the output timing, while spaced by T, could be offset in phase such as $4.1T$, $5.1T$, ..., and so on. This can happen in several ways: if the system has a filter with group delay, $\hat{o}_g$, which is not a multiple of T, or artifacts may be associated with the processing delay of various operations.

The FEC is again a good example. Consider a system with $T = 1$, using an triple error correcting FEC with $k = 12$ and $N = 23$. So an input and output block is 12 seconds long. But the output bits are spaced 12/23 seconds apart. From Chapter 5 we saw that to detect an NRZ data bit, an integrate and dump matched filter with integration time 12/23 seconds long is required. Thus, the start of the code word at the receiver is shifted by this amount, and the output bits, will occur at times $T = 12/23$, 1 12/23, 2 12/23. The output data bit stream will be shifted by the same amount. One solution to this problem is to relabel the output time axis to match the input time axis. These idiosyncrasies must be carefully observed if the simulation is to be a success.

9.3   E$_b$/N$_0$ Calibration

The nearly universal independent variable for measuring BER is the energy per bit, as described in Chapter 5. We now turn to how to calibrate this parameter in a simulation. There is more than one way, of course, but the method presented here is easy to understand and implement. There are two basic simulation types to consider here.

9.3.1   Physical RF World Simulation

In this type of simulation, exact account is taken of the real physical power levels of the system as well as the physical noise levels encountered at every step in the reciever. In short, you want to know what power level (in dBm) is required to meet a certain goal. From link equations, this power level translates to communication range. Figure 9.4 shows a typical block diagram of such a simulation.

Start with the white Gaussian noise power density N_0 (W/Hz). From thermodynamic considerations, we know that $N_0 = kT$, where $k = 1.38 \times 10^{-23}$ W/HzK is Boltzman's constant, and T is a suitable chosen reference temperature commonly $T = 290K$ (to make kT a nice round number). This gives

$$N_0 = \left(1.38 \cdot 10^{-23} \ W/Hz \ K\right)(290K)$$
$$= 4 \cdot 10^{-21} \ W/Hz$$
$$= 4 \cdot 10^{-18} \ mW/Hz$$
$$= -174 \, dBm/Hz$$

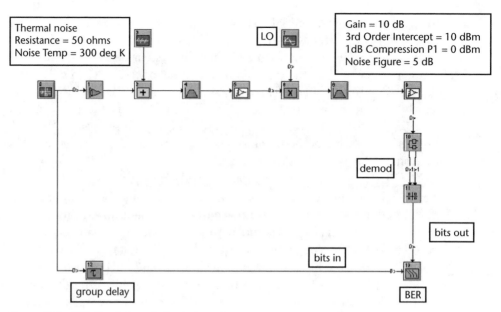

Figure 9.4 BER Simulation of physical receiver.

The last representation is good to remember as it is frequently used in system sensitivity and noise figure calculations.

Generating a Gaussian noise source with a specific N_0 is accomplished by the relation

$$N_0 R f_s / 2 = \sigma^2$$

where fs is the system simulation sample rate, σ^2 is the variance of the Gaussian noise source, and R is the resistance of the power measurement system, usually 50 ohms. This result is called Parceval's theorem, which states that the power of the signal in the time domain is equal to the power in the frequency domain. When calibrating the SNR, be sure that both the noise and power are measured with respect to the *same R*.

The average power transmitted P_{av} is calculated as

$$P_{av} = \int_0^T s^2(t)\, dt / T$$

where $s(t)$ is the final modulated carrier signal. The measurement is made just before any noise or interference is added. The integration time T is long enough to ensure an accurate result. This operation needs only to be performed once when the simulation is being set up.

In some cases, the calculation of P_{av} can be done almost by inspection. Any constant envelope signal, of amplitude A (zero-to-peak) (such as MFSK), a pure tone, or MPSK without any baseband filters, will have

$$P_{av} = A^2 / 2R$$

With the signal and noise calibrated, at some point the BER calculation can be made. The power level is controlled to some value for each run, and the BER is calculated. A plot is then made of BER versus P_{av}.

9.3.2 Standard BER Runs

All digital communication formats such as BPSK and QAM have well-known performance characteristics. They are plotted BER versus the energy per bit E_b/N_0. Recall from Chapter 5 that the output of a matched filter is $2E_b/N_0$, so these plots essentially give a baseline performance. It is not uncommon to have a system with a specification, for example, that QPSK modulation shall be employed, and the performance is to be within 1 dB of the optimum performance curve. So what we need is to set up the simulation parameters for the noise accordingly. The basic relation is

$$E_b/N_0 = P_{av}/N_0 R_b$$

where R_b is the basic system bit rate (not the symbol rate). P_{av} is the same average power as discussed in Section 9.3.1:

$$P_{av} = \int_0^T \left[I^2(t) + Q^2(t) \right] dt \Big/ T$$

Now that we have an initial calibration, we can vary E_b/N_0 to any desired value in one of two ways. We can increase the signal power by inserting a gain element calibrated in decibel power right after the signal $s(t)$ is formed. We can also decrease the noise power in a similar way, only with the gain value being negative decibels. The author prefers the second method. Many elements in a system may be amplitude sensitive. If there are nonlinear elements, then the spurs will increase with increased power. The bandwidth (time response) of a phase locked loop is also input gain sensitive. Either method works, so you can choose the one best suited to your needs.

9.4 Conclusion

This chapter was devoted to the important concept of BER measurements. The BER is the near universal figure of merit of a communications link. The main issues were the calibration of the energy per bit E_b/N_0, and establishing the delay through the system.

Selected Bibliography

Pahlavan, K., and A. H. Leveseque, *Wireless Information Networks*, New York: John Wiley & Sons, 1995.

Proakis, J. B., *Digital Communications*, New York: McGraw-Hill 1983.

Rappaport, T. S., *Wireless Communications, Principles and Practice,* Upper Saddle River, NJ: Prentice Hall, 2002.

Sklar, B., *Digital Communications, Fundamentals and Applications,* Upper Saddle River, NJ: Prentice Hall, 2001.

Steele, R., *Mobile Radio Communications,* London: Pentech Press Limited, 1992.

CHAPTER 10

Channel Models

In Chapter 1 we defined the channel to be the media that the signal passes through from the transmit antenna to the receive antennas. One major component that describes the channel is the signal fading phenomena. As shown in Figure 10.1, the transmitted signal can bounce off objects, each reflecting energy to the receive antenna. The sum total of all of these reflections can cause the resulting signal to lose relative power by cancellation, either in total or at select frequencies. We now develop in more detail this concept and some methods of simulation.

10.1 Flat Fading

Consider a transmitted signal of the form

$$s(t) = A(t)\sin\left[2\pi f_0 + \varphi(t)\right]$$

Now suppose that this signal bounces off many object and reflects back to the receiver. Each path has a delay τ. The difference between the maximum and minimum delay is called the delay spread. The net received signal $r(t)$ is

$$r(t) = \sum_k s(t - \tau_k) = \sum_k A_k(t - \tau_k)\sin\left[2\pi f_0 t + \varphi(t - \tau_k) - 2\pi f_0 \tau_k\right]$$

The concept of flat fading is based on the following observation. If a signal is simply passed through a delay line, then the output can be written as

$$s_d(t) = \int_{-\infty}^{\infty} \left[S(f)e^{-2\pi jf\tau}\right]e^{2\pi jft}\, df$$

Now if the spectrum $S(f)$ is confined to a bandwidth $B \ll 1/\tau$, the exponential term in the bracket is essentially constant over the active integration, giving

$$s_d(t) = \int_{-B}^{B} \left[S(f)e^{-2\pi jf\tau}\right]e^{2\pi jft}\, df$$

$$= \int_{-B}^{B} \left[S(f)e^{-2\pi jf\tau}\right]e^{2\pi jft}\, df$$

$$= \int_{-B}^{B} \left[S(f)\cdot 1\right]e^{2\pi jft}\, df$$

$$= s(t)$$

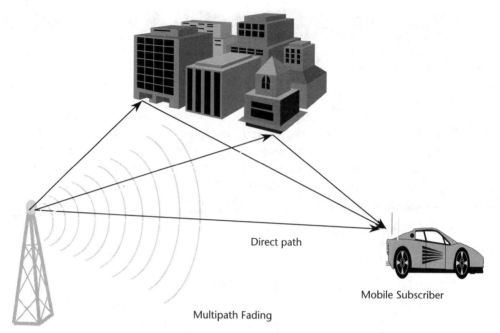

Figure 10.1 General model for multipath fading. The signal can bounce off of numerous objects and recombine at the receiver.

For all practical purposes, the delay does not affect the modulation component of the signal except for a possible gain factor α_k. With this in mind, we can rewrite the fading signal as

$$r(t) = A(t)\left[\sin(2\pi f_0 t + \varphi)\sum_k \alpha_k \cos(2\pi f_0 \tau_k) - \cos(2\pi f_0 t + \varphi)\sum_k \alpha_k \sin(2\pi f_0 \tau_k)\right]$$

The trick now is to evoke the law of large numbers to the summation terms. In particular, the PDF of a sum of independent variables becomes Gaussian in the limit of a large number of terms, regardless of the PDF of the individual components.

The final expression for the fading signal now becomes

$$r(t) = A(t)\left[a(t)\sin(2\pi f_0 t + \varphi(t)) + b(t)\cos(2\pi f_0 t + \varphi(t))\right]$$
$$= A(t)M(t)\sin(2\pi f_0 t + \varphi(t) + \theta(t))$$

$$M(t) = \sqrt{a(t)^2 + b(t)^2}$$
$$\theta(t) = \tan^{-1}\left[a(t)/b(t)\right]$$

where $a(t)$ and $b(t)$ are zero mean Gaussian signals.

So the net result of the fade is to amplitude modulate the signal with $M(t)$ and introduce phase shift $\theta(t)$. The term "flat" fading now refers to the fact that the whole of the signal breathes up and down like a time varying gain function. Such a function acts equally (flat) on all frequency components of the original signal.

Owing to the fact that $a(t)$ and $b(t)$ are Gaussian variables, the amplitude distribution of $M(t)$ is calculated to be

$$P(x,y) = e^{-(x^2+y^2)/2\sigma^2}/2\pi\sigma^2$$

$$P(r,\theta) = e^{-r^2/2\sigma^2}/2\pi\sigma^2$$

$$P(r) = \int_0^{2\pi} P(r,\theta)\,rd\theta$$

$$= re^{-r^2/2\sigma^2}/\sigma^2$$

This function is the well-known Rayleigh distribution. The distribution of the phase angle $\theta(t)$ is uniform $[0, 2\pi]$.

Simulation of flat fading depends on whether the simulation is baseband, or RF. Figure 10.2 shows the baseband case.

In Figure 10.2 the Gaussian noise is filtered to some bandwidth B to control the fade rate.

Simulating a flat fade when using an RF simulation (i.e., the signal is on a carrier) is somewhat more complicated. Remember, the model affects both the amplitude and the phase of the signal. If the signal is of the form

$$s_i(t) = A(t)\sin\left[2\pi f_0 t + \varphi(t)\right]$$

we need to generate the quadrature signal

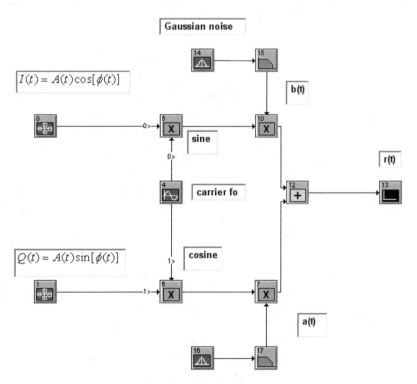

Figure 10.2   Simulation of a fading channel at baseband.

$$s_q(t) = A(t)\cos\left[2\pi f_0 t + \varphi(t)\right]$$

in order to proceed. Figure 10.3 shows the overall process.

The Hilbert filter is the standard method for producing the quadrature of a signal. Essentially it shifts all frequencies by 90 degrees. The delay τ is required to match the group delay introduced by the filter.

10.2 Rician Flat Fade

In the Rayleigh fade model, all of the energy reaching the receiver is the result of some bounce. There is no direct transmitter to receiver path. The Ricean model adds this direct path. Figure 10.1 shows this model, where there is a direct path and a series of bounces from the base station to the mobile subscriber. The amplitude distribution is a little more complicated now:

$$P(x,y) = e^{-\left[(x-A)^2 + y^2\right]/2\sigma^2}\big/2\pi\sigma^2$$

$$P(r,\theta) = e^{-\left[r^2 + A^2 - 2rA\cos\theta\right]/2\sigma^2}\big/2\pi\sigma^2$$

$$P(r) = e^{-\left[r^2 + A^2\right]/2\sigma^2}\int_0^{2\pi} e^{2rA\cos\theta/2\sigma^2}\, rd\theta/2\pi\sigma^2$$

$$= re^{-\left[r^2 + A^2\right]/2\sigma^2} I_0\left(Ar/\sigma^2\right)\big/2\pi\sigma^2$$

where $I_0(x)$ is the modified Bessel function of order 0, and A is the amplitude of the direct path. In the limit that A goes to 0, we recover the original Rayleigh distribu-

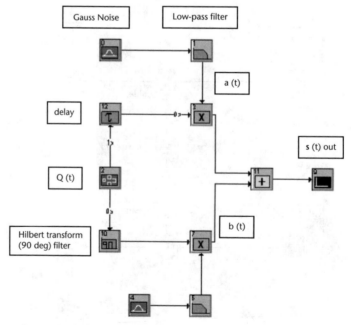

Figure 10.3 Implementing a Rayleigh fading channel at RF.

tion. In this model the parameter $K = (A/\sigma)$ is used to normalize the ratio of the direct to reflected power. $K = 0$ is a pure Rayleigh channel, and $K = \infty$ is a pure direct path.

10.3 The Jakes' Model

So far, the time dynamics of the channel have been represented by simply filtering Gaussian noise to the appropriate bandwidth. We now turn to the simple and often used model, from Jakes, shown in Figure 10.4.

In this model a mobile subscriber is traveling with a velocity v with respect to the base station tower. It is assumed that there is a large number of scattering paths that reach the mobile unit from all directions, as shown. Now the motion of the received path causes a Doppler shift depending on the angle of arrival θ with respect to the motion. The Doppler shift of a signal is given by the equation

$$f_d = f_0 v \cos(\theta)/c$$

From this, we see that $f_d = 0$ for arrival from above and below, $f_d = -f_0 v/c$ for arrival from behind, and $f_d = +f_0 v/c$ for arrival from the front. The range of frequencies in a mobile channel is called the Doppler spread. The PDF of the Doppler spread in this case is v given by

$$S(f) = 1\Big/ \sqrt{1 - (f/f_d)^2}$$
$$f_d = v f_0/c$$

The issue now is how to generate a fading signal [I, Q] that has this distribution. A commonly used model, from Jake, is shown in the following equations:

$$I(t) = 2\sum_{n=1}^{N} \cos\varphi_n \cos\omega_n t + \sqrt{2} \cos\varphi_N \cos\omega_m t$$

$$Q(t) = 2\sum_{n=1}^{N} \sin\varphi_n \cos\omega_n t + \sqrt{2} \sin\varphi_N \cos\omega_m t$$

$$\omega_n = \omega_m \cos(2\pi n/N)$$
$$\omega_m = 2\pi f_m = 2\pi f_0 v/c$$

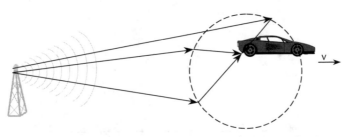

Figure 10.4 The Jakes' mobile multipath model. It is assumed that the reflection paths are equally distributed from a circle around the receiver.

The larger the number of terms, N, the better the representation will be. A value of $N = 10$ or so is generally sufficient. Figure 10.5 shows the actual $S(f)$ obtained with $N = 10$, and Figure 10.6 shows the corresponding Rayleigh amplitude PDF .

The result is a good approximation to the Rayleigh distribution.

Observe that the [I, Q] of this representation is purely deterministic. There is no random variable parameter. One way to fix this is to add a random phase to each of the sinusoidal terms.

10.4 Frequency Selective Fading

Consider a channel with a direct and single reflective fade path that inverts the signal. The difference in the time of arrival of these two paths is τ. The channel model in the frequency domain is simply

$$h(t) = 1 - \delta(t - \tau)$$
$$H(f) = 1 - e^{2\pi jft}$$

Clearly, if $f\tau$ is an integer, there is complete cancellation. Figure 10.7 shows the result of this action on a PSK signal spectrum.

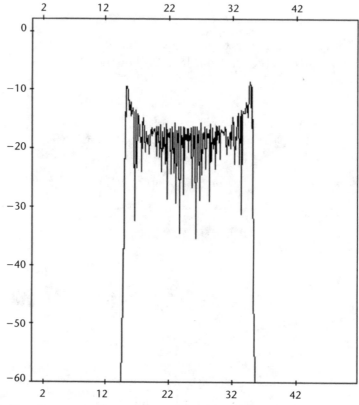

Figure 10.5 Doppler spectrum from Jake's model using $N = 10$ terms.

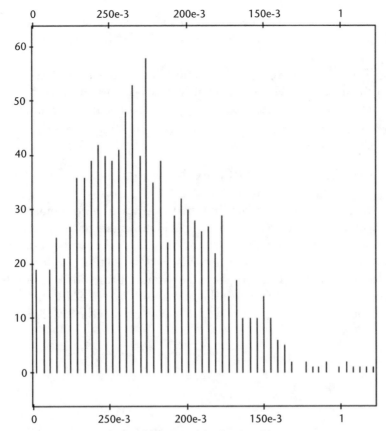

Figure 10.6   Amplitude PSD obtained from model using *N* = 10 terms.

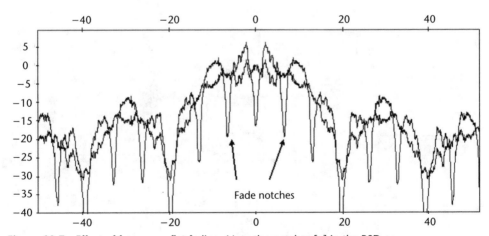

Figure 10.7   Effect of frequency flat fading. Note the notches [x] in the PSD.

The data rate is 10 Hz, and the path delay is 0.15 sec. The frequency cancellations occur at 0, 6.67, 13.33 Hz, as shown in Figure 10.7. This notch phenomenon gives rise to the term frequency selective fade.

The model for frequency selective fade is based on a tapped delay line structure, shown in Figure 10.8. The basic idea is to combine several frequency flat channels separated by some time difference.

The gain in each arm is actually represented by frequency flat dynamics as already described. The wireless industry has performed extensive field tests to arrive at realistic fade models. Table 10.1 shows the standard model cited in the IS-95 CDMA specification.

10.5 Effects of Fading on BER Performance

How does fading affect the BER performance of a communication channel? The received signal power is breathing up and down due to the fading. In a slow fade situation (i.e., one where the fade rate is less than the symbol rate), one can consider the received signal to be constant over the symbol time. In this situation we can calculate the resultant BER by taking the formula already derived for BER as a function of SNR, and weight it against the probability of that SNR.

For the BPSK case, we showed in Chapter 5 that the BER is given by the formula

$$P(e) = Q\left[\sqrt{2E_b/N_0}\right] = \int_{\sqrt{2E_b/N_0}}^{\infty} e^{-\mu^2/2} \, du/\sqrt{2\pi}$$

Note that $E_b = A^2/2R$. For a channel, the PDF of the amplitude A is given by

$$P(A) = \left(A/\overline{A^2}\right)e^{-\left(A^2/2\overline{A^2}\right)}$$

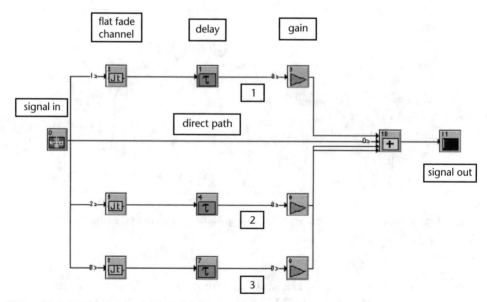

Figure 10.8 Tapped delay line model of a frequency selective fade channel. The gain token in each path represents the dynamics of a frequency flat fade channel.

Table 10.1   IS-95 Cell Phone Channel Models

Model	Vehicle Speed (km/hr)	Number of Paths	Path 2 Power Relative to Path 1 (dB)	Path 3 Power Relative to Path 1 (dB)	Path 2 Delay Relative to Path 1 (usec)	Path 3 Delay Relative to Path 1 (usec)
1	8	2	0	N/A	3.0	N/A
2	30	1	N/A	N/A	N/A	N/A
3	100	3	0	−3	2.0	14.5

where $\overline{A^2}$ is the average of the square of the signal amplitude. The expression for the fade rate P_{fade} of the BPSK channel is the weighted integral P(e) (which is a function of A), against P(A)

$$P_{fade} = \int_0^\infty P(e)P(A)dA$$

$$P_{fade} = \int_0^\infty \left(A/\overline{A^2} \right) e^{-\left(A^2/2\overline{A^2} \right)} dA \int_{\sqrt{A^2/N_0 R}}^\infty e^{-u^2/2} \, du/2\pi$$

This integral can be evaluated:

$$P_{fade} = .5\left\{ 1 - \left[\overline{SNR}/\left(1 + \overline{SNR} \right) \right]^{.5} \right\}$$

$$\overline{SNR} = \overline{A^2}/2N_0 R$$

Figure 10.9 plots this result against the nonfade BER.

Table 10.2 shows the results for this and several other cases in the limit of high $\overline{SNR}$.

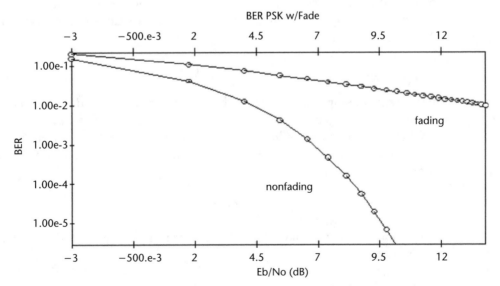

Figure 10.9   BER of a BPSK channel with and without fading. Note the severe performance penalty of the fading case.

Table 10.2 BER Performance in
Rayleigh Fading Channel

Modulation Type	BER
Coherent PSK	$1/4\overline{SNR}$
Coherent FSK	$1/2\overline{SNR}$
DPSK	$1/2\overline{SNR}$
Noncoherent FSK	$1/\overline{SNR}$

Note this very important observation: The performance difference for coherent versus noncoherent transmission is a factor of two. This is one reason that in the IS-95 wireless system, the base station transmits a signal that allows the receiver to establish absolute phase reference permitting coherent reception. This is economical in this case since it is a one-to-many transmission format.

10.6   Mitigating Fading

Given that fading is a bad thing, how do we mitigate its effects? There are two general methods.

10.6.1   Equalizers

We just described the general channel model in the mathematical format

$$h(t) = \sum_k c_k(t)\delta\big[t - \tau_k(t)\big]$$

where the coefficients $c_k(t)$ represent the Rayleigh fading, and $\tau_k(t)$ represents the various path delays, which may be a function of time as well. Now, over a short enough time span, these dynamic quantities can be considered constant. In this case the channel is simply a filter. In the digital z domain, we can write the transfer function across this filter as

$$Y(z) = H(z)X(z)$$

where $X(z)$ is the input data, $H(z)$ is the channel transfer function, and $Y(z)$ is the resulting received signal.

What we want to do is insert a filter with some transfer function $F(z)$ at the receiver in order to mitigate the effects of the channel. The received waveform after this processing is simply

$$Y(z) = F(z)H(z)X(z)$$

The objective is to make $Y(z) = X(z)$, which implies that we choose this filter such that

$$F(z)H(z) = 1$$

A filter $F(z)$ that accomplishes this goal is called an equalizer.

The basic form of a linear equalizer is taken as an FIR filter structure of the form

$$F(z) = \sum_k a_k z^{-k}$$

and the whole of equalizer theory is how to choose the coefficients a_k.

As an example, Figure 10.10 shows a block diagram of a seven-tap linear equalizer. The transmitted signal is binary ±1, which has been slightly filtered. The transmission channel has a simple transfer function $H(z) = 1 - 0.5z^{-5}$. The system simulation rate is set to be five times the data rate. Thus, for simplicity, the channel delays by 1 data bit and adds back half of the value seen.

This operation introduces ISI into the system. In Chapter 8 we showed that the eye diagram was a convenient method for investigating this phenomenon. Figure 10.11 shows the eye diagram of the signal after the channel $H(z)$. Note that the eye has four levels and is nearly closed.

In this example of Figure 10.10, the error signal e_k at sample time t_k is derived by comparing the actual signal y_k with the nearest possible actual data value equal to ±1. Thus the hard limiter is the proper mathematical operation. In a more complicated system, such as MPSK, the signal, coefficients, and the error signals are all complex entities. The error signal is then the difference between the instantaneous I/Q values and the nearest member of the signal constellation.

The remaining problem is the computational algorithm that calculates the taps. We have chosen a very well-known and simple method known as the LMS algorithm, introduced by Widrow. The block diagram of the taps of Figure 10.10, which implement this algorithm, is shown in Figure 10.12.

The mathematical calculation is given by the vector formula

$$\left[\hat{a}\right]_{k+1} = \left[\hat{a}\right]_k + 2\mu\left[e_k\right]\left[\hat{x}\right]_k$$

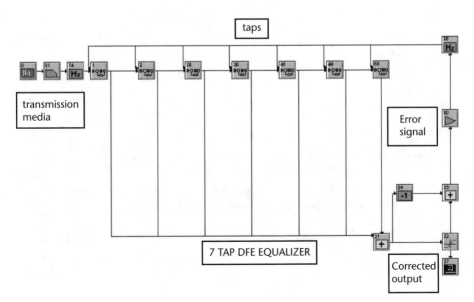

Figure 10.10 Block diagram of a seven-tap linear equalizer.

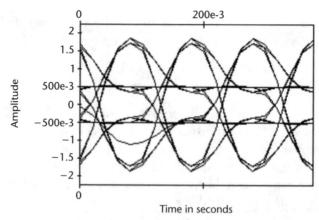

Figure 10.11 Eye diagram of BPSK signal after channel. The eye is essentially closed.

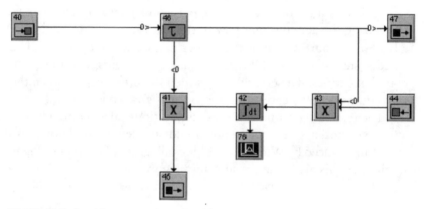

Figure 10.12 LMS algorithm tap update algorithm.

where $[\hat{a}]_k =$ is the array of tap coefficients at step k; $[\hat{a}]_k =$ is the array of tap coefficients at step $k + 1$; $[\hat{x}]_k =$ is the array of input signals at step k; e_k is error at step $k =$ [desired − actual] $_k$; and μ is the loop gain constant.

As noted, the perfect equalizer will produce a filter of the form

$$F(z) = 1/\left(1 - z^{-5}\right) = 1 + 5z^{-5} + 25z^{-10} + \ldots$$

The first thing to note is that it would take an infinite number of taps to realize this filter; thus, the seven-tap version here will only correct the channel to some point. Figure 10.13 shows the evolution of the taps as the algorithm converges. Table 10.3 summarizes the theoretical versus actual coefficients for the first three taps.

Finally, Figure 10.14 shows the eye diagram after the equalization.

There are several other equalizer structures in addition to the LMS example given above.

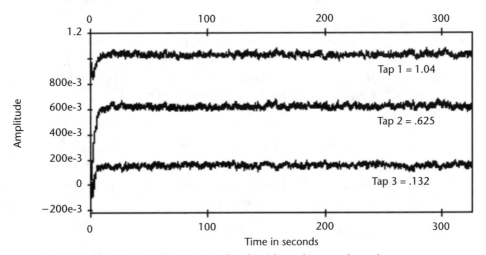

Figure 10.13   Evolution of equalizer taps as the algorithm adapts to channel.

Table 10.3   The Theoretical Versus Actual Coefficients

Tap number	Perfect	Seven-tap model
1	1	1.04
2	0.5	0.625
3	0.25	0.132

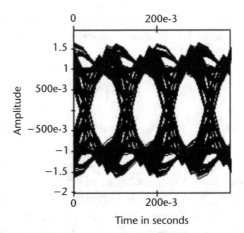

Figure 10.14   Eye diagram after the seven-tap equalizer. The eye has opened up.

CMA Algorithm

This system works on any constant envelope or modulus (power level = $I^2 + Q^2$ = constant) signal such as MPSK and MFSK. The error in Figure 10.11 will simply be the difference between the actual power level and a fixed reference.

Decision Feedback Equalizers

In the transversal structure of Figure 10.11, there is no coupling of the data decisions back into the equalizer algorithm. In the decision feedback equalizer (DFE) the data decisions are fed back into the equalizer, and second set of weights is used to improve the performance. In effect, the equalizer now has the z domain representation of

$$H(z) = N(z)/D(z)$$

where the numerator $N(z)$ is derived from the transversal portion, and $D(z)$ is derived from the feedback portion of the equalizer.

Zero Forcing Equalizer

As noted above, the effect of the channel is to introduce ISI into the detected symbols. Now suppose we transmit a signal with a single raised cosine pulse $p(t)$ described in Chapter 8. We saw that this filter introduces no ISI since $p(kT) = 0$ at all multiples of the data rate except at the center $k = 0$. At the receiver we sample the received signal $x(t)$ at time kT, X_k. Because of the channel ISI, X_k is generally not equal to zero. The idea of the zero forcing equalizer (ZFE) is to choose the tap weights that enforce the original zero ISI condition. For a three-tap ZFE, the equations are

$$\begin{bmatrix} y_{-1} \\ y_0 \\ y_1 \end{bmatrix} = \begin{bmatrix} 0 \\ 1 \\ 0 \end{bmatrix} = \begin{bmatrix} x_{-2} & x_{-1} & x_0 \\ x_{-1} & x_0 & x_1 \\ x_0 & x_1 & x_2 \end{bmatrix} \begin{bmatrix} a_{-1} \\ a_0 \\ a_1 \end{bmatrix}$$

This is a system of three equations in three unknowns, which can be solved for the tap weight vector [a]. In matrix notation we have

$$[I] = [X][a]$$
$$[a] = [X]^{-1}[I]$$

The more taps used, the greater the accuracy of the filter.

Viterbi Equalizer

The Viterbi equalizer (VBE) is employed on the GSM/GMSK mobile cell phone system. For severe interference environments such as cities, the fading occurred can be quite significant. In addition, the dynamic environment due to the mobile receiver motion requires constant updating of the channel. The GSM messages are configured into slots 0.577 msec long containing 156.25 bits. In each slot there is a 26-bit training sequence $h(t)$, and two groups of 58-bit data sequences, plus guard bits and nonmodulated guard time. If we denote the transfer function of the channel by $H(t)$, then the received training signal $r(t)$ is simply the convolution

$$r(t) = H(t) \otimes h(t)$$

At the receiver we correlate $r(t)$ with $b(t)$ to get the signal $z(t)$

$$z(t) = r(t) \otimes b(t)$$
$$= H(t) \otimes \big[b(t) \otimes b(t) \big]$$

If we carefully choose $b(t)$ such that $[b(t) \otimes b(t)] \approx \delta(t)$, then $z(t)$ is an estimate of the channel

$$z(t) \approx H(t)$$

The next step is to correlate all possible bit combinations that are correlated with $z(t)$. The receiver then compares the actual data sequence with this set in a trellis decoder similar to the one used for the Viterbi error correcting algorithm detailed in Appendix C.

10.6.2 Diversity

The basic idea of diversity is to get more than one "look" at the received signal. Hopefully, each look will not fade at the same time. By combing these paths, a reliable channel is established. There are many possible diversity concepts:

- *Spatial:* If we have two antennas placed further apart than the carrier wavelength, then the complex cancellation phenomena will be different for each antenna position. For a mobile system, one antenna could be on the front of the car, and the other on the rear.
- *Polarization:* It may be possible to use two orthogonal polarization receive antennas: horizontal and vertical. They resulting signals may fade independently.
- *Time:* The individual paths of the delay line model have a different delay. If we could resolve out these delays, then each path could be detected individually and then combined.
- *Frequency:* The fade at one frequency may be entirely different than the fade at another.

One of the primary uses of spread spectrum modulation techniques is to accomplish this goal. Recall that the correlation peak of a DSPN system is one chip wide. Thus, if the received delay paths are greater than this chip period, they can be individually demodulated. This is exactly the tack taken in the IS-95 CDMA wireless system. Figure 10.15 shows a three-path block diagram of such a system. The general term of this algorithm is a *rake receiver*. Each path is a finger of the rake.

What is the performance gain from the diversity? An analysis shows that for L diversity paths, the probability of error is given by

$$P_L(e) \approx \big(P_1(e) \big)^L$$

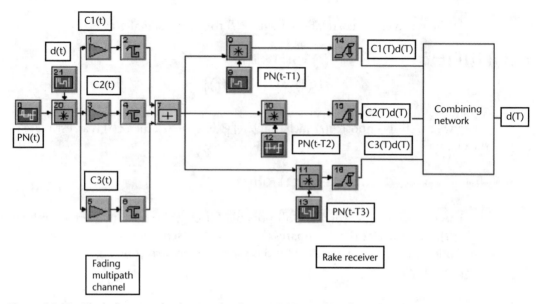

Figure 10.15 Block diagram of a three-path rake type DSPN multipath receiver.

In the case of the three-finger rake system used by IS-95 BER out for a $P_1 = 0.1$ we get $P_3 = 0.001$. Furthermore, with coherent reception the gain is a factor of eight better than for noncoherent transmission.

10.7   Conclusion

This chapter was devoted to what happens to the signal between the transmitter and receiver. This medium is commonly called the channel. In modern wireless systems, the fading nature of the channel dominates performance over the effects of AWGN.

Selected Bibliography

Jakes, W. C., (Ed.), *Microwave Mobile Communications,* New York: John Wiley & Sons, 1974.

Pahlavan, K., and A. H. Leveseque, *Wireless Information Networks,* New York: John Wiley & Sons, 1995.

Proakis, J. B., *Digital Communications,* New York: McGraw-Hill 1983.

Rappaport, T. S., *Wireless Communications, Principles and Practice,* Upper Saddle River, NJ: Prentice Hall, 2002.

Sklar, B., *Digital Communications, Fundamentals and Applications,* Upper Saddle River, NJ: Prentice Hall, 2001.

Steele, R., *Mobile Radio Communications,* London: Pentech Press Limited, 1992.

Widrow, B., and S. D. Stearns, *Adaptive Signal Processing,* Englewood Cliffs, NJ: Prentice Hall, 1985.

CHAPTER 11

Nonlinear Amplifiers

In this chapter we describe methods for simulating the nonlinear behavior of RF amplifiers. We include the traveling wave tube (TWT) amplifier commonly used on communication satellites, and model a mixer spur chart.

11.1 Intercept and Compression Points

The simplest model of a nonlinear amplifier is given by the transfer function

$$y = ax + cx^3$$

where a is the linear gain term. The standard specification of the nonlinearity is called the third order two-tone intercept point, IP_3. To see what this means, we take the input signal x as the sum of two equal amplitude sinusoids (tones)

$$x = A\sin(2\pi f_1 t) + A\sin(2\pi f_2 t)$$

and see what y looks like. The result is

$$
\begin{aligned}
y &= a\big[A\sin(2\pi f_1 t) + A\sin(2\pi f_2 t)\big] \\
&+ c\big[A\sin(2\pi f_1 t) + A\sin(2\pi f_2 t)\big]^3 \\
&= aA\big[\sin(2\pi f_1 t) + \sin(2\pi f_2 t)\big] \\
&+ cA^3\big[\sin^3(2\pi f_1 t) + \sin^3(2\pi f_2 t)\big] \\
&+ 3cA^3\big[\sin^2(2\pi f_1 t)\sin(2\pi f_2 t) + \sin(2\pi f_1 t)\sin^2(2\pi f_2 t)\big]
\end{aligned}
$$

Now we substitute the three standard trigonometric identities:

$$
\begin{aligned}
\sin^3\alpha &= \big[-\sin(3\alpha) + 3\sin(\alpha)\big]/4 \\
\sin^2\alpha &= \big[1 - \cos(2\alpha)\big]/2 \\
\sin\alpha\sin\beta &= \big[\cos(\alpha - \beta) - \cos(\alpha + \beta)\big]/2
\end{aligned}
$$

then gather terms of the same frequency. The result is summarized in Table 11.1. The following notation applies: $x = 2\pi f_1 t$, and $y = 2\pi f_2 t$. In addition, we used frequencies $f_1 = 10$ Hz and $f_2 = 11$ Hz to make things more specific.

Table 11.1   Nonlinear Harmonic Coefficients

Coefficient	Harmonic Term	Value (Hz)
$aA + 3cA^3/4$	$\sin(x)$	10
$aA + 3cA^3/4$	$\sin(y)$	11
$-3cA^3/4$	$\sin(3x)$	30
$3cA^3/4$	$\sin(3y)$	33
$3cA^3/2$	$\cos(2x - y)$	9
$3cA^3/2$	$\cos(2y - x)$	12
$-3cA^3/2$	$\cos(2x + y)$	31
$-3cA^3/2$	$\cos(2y + y)$	32

In the general case with higher order nonlinearities, the procedure shown above will eventually produce trigonometric terms of the form $\sin(mx + my)$. The harmonic order l is given by $l = m + n$.

The four third order harmonic terms in the 30-Hz range are usually eliminated by system filters. Note, however, that there are two third order terms, one at 9 Hz and the other at 12 Hz, that are in the direct band of 10 Hz and 11 Hz.

Note in the table that the amplitude of the $\sin(x)$ gain term increases as A, while the amplitude of the third order terms like $\cos(2x - y)$ increases as A^3. So as A increases there will be a point where the two power levels are the same. This point is called the third order intercept point IP_3, from which we have

$$IP_3 = (aA)^2 / 2R$$

$$IP_3 = (3cA^3/4)^2 / 2R$$

where R is the resistance that the power is dissipated into, nominally 50 ohms for most systems. From these two equations we can solve for c in terms of IP_3 and a, alone. The result is

$$c = \pm 2a^3 / 3IP_3 R$$

Choosing between the + or − sign is not arbitrary. Either will work as far as the power, but review the coefficient of the $\sin x$ term in Table 11.1. If $c > 0$, then as the input amplitude A increases, the net power would increase above the linear gain term. This is not physically correct. If $c < 0$, the output power will begin to fall below the extended linear projection. This is the physical result. The amplifier cannot put out infinite power as the input power increases. It will go into saturation where the output levels off to some value regardless of the input. By taking $c < 0$, we achieve this result.

Example

Design an amplifier model that has a 10-dB linear gain and an IP_3 of +10 dBm.

Solution: From the above equations, we arrive at the transfer equation

$$y = \sqrt{10}x - 42.16x^3$$

Figure 11.1 shows the signal spectra when the input signal power is –20 dBm. From this figure, the output power of the direct signal is 10 dB higher than the input, or –10 dBm. The third harmonic power is at -50 dBm. Now if we increase the input power by 20 dB, the direct power increases in kind to +10 dBm. But the third harmonic term increases as a 3:1 ratio, so its output power is now –50 dBm + 3*20 dBm = +10 dBm again. This verifies the model.

The model is not quite finished, however. As the input keeps increasing the input power, the output power will reach some maximum, and then start to decrease again. The curve is a cubic. What we can do is find the point where the curve maximizes out. This is given at the input amplitude

$$A_{max}^2 = 2a/3c$$
$$P_{max} = A_{max}^2/4$$

At this point the curve is at its maximum and the slope is horizontal. If the input power is greater that A_{max}, we simply extend the P_{max} out to infinity. This gives a smooth transition and is quite satisfactory in terms of performance. Figure 11.2 shows this model.

The 1-dB compression point is a standard amplifier specification, as is the IP_3. This point is defined as the output power that is 1 dB below that which would be expected for a linear amplifier.

Note that we do not have independent control over the 1-dB compression point. But it will fall within the desired range mention above. In this model, Figure 11.3 shows that the 1-dB point is +0.3 dB.

In general, the IP_3 is about 10 to 12 dB higher than the P_1 point. The amplifier will go into saturation before the IP_3 point is reached. In the model used here we have such a result. The IP_3 is slightly less that 10 dB greater to than the P1. Note that this result is fixed. We cannot independently control both parameters with this simple third order model. In order to independently control the 1-dB point, we must use a higher order polynomial to describe the curve.

Figure 11.4 shows the standard plot that combines both effects.

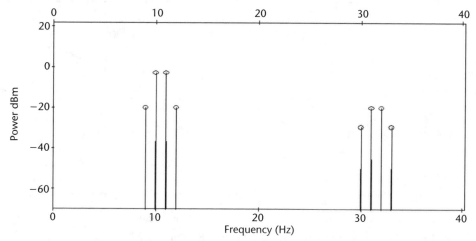

Figure 11.1   Signal frequencies and spurs arising from a third order nonlinearity.

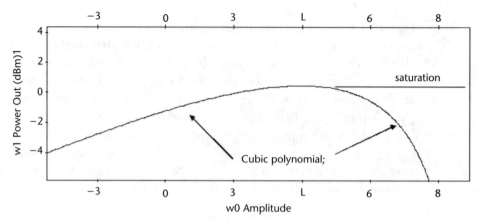

Figure 11.2 The cubic model with the saturation extension from the maximum point of the cubic polynomial.

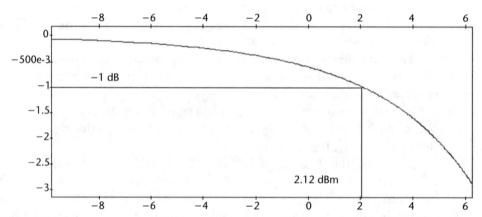

Figure 11.3 The output power divided by the gain as a function of the input power. The indicated point is the P_1.

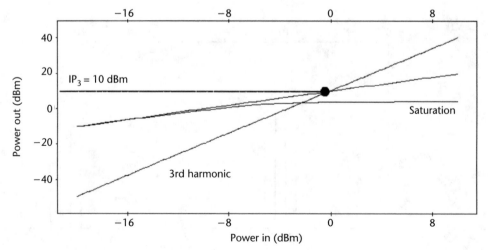

Figure 11.4 Combined performance plot of a nonlinear amplifier.

11.2 TWT

The TWT is also a nonlinear amplifier, but it is described differently than above. The basic equation is

$$y(r,t) = A(r)\sin(2\pi ft + \varphi(r))$$

where r is the amplitude of the input signal. The term $A(r)$ is known as AM-AM conversion, and the term $\varphi(r)$ is known as AM-PM conversion. One model of a TWT is

$$A(r)/r = a_r/(1+b_r r^2)$$
$$\varphi(r) = a_\varphi r^2/(1+b_\varphi r^2)$$

where the four coefficients are found via a curve fit to the actual measured data. A simpler way is to use a table look-up format. Figure 11.5 shows the plot for the amplitude, and Figurer 11.6 shows the plot for the phase.

An easy implementation of the model is at baseband where the phase and amplitude can be controlled independently. This model is shown in Figure 11.7.

If the model is used at RF, then a Hilbert transform must be used to produce a quadrature signal that is required to implement the phase shift. This model is shown in Figure 11.8.

11.3 Spur Charts

Another standard presentation of the amplifier nonlinearity is a spur chart shown in Table 11.2. This chart usually goes up to a 10 × 10 matrix. It shows the power of the harmonic $m \times n$ relative to the direct output.

Theses charts are always presented under some specific conditions as shown in the table. The [1, 1] position entry in the matrix is a "−". It can be interpreted as 0 dBm (relative to itself), which is the notation sometimes seen.

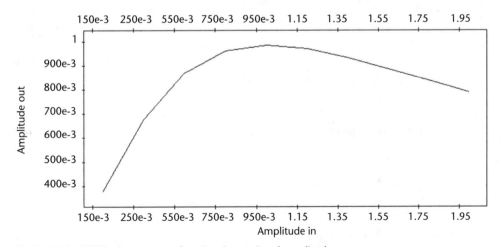

Figure 11.5 TWT tube gain as a function input signal amplitude.

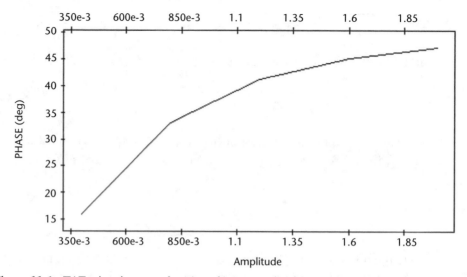

Figure 11.6   TWT tube phase as a function of input amplitude.

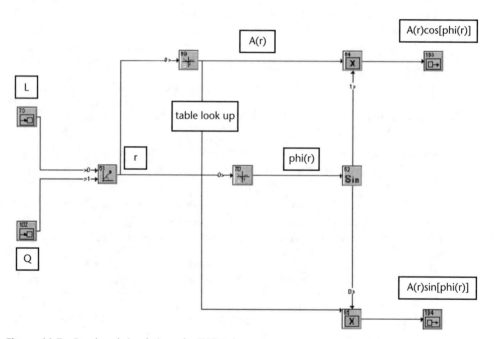

Figure 11.7   Baseband simulation of a TWT tube.

One way to model this chart is to write the amplifier transfer function in the form

$$y = \sum_{k=0}^{10} a_k x^k$$

Next, substitute for x

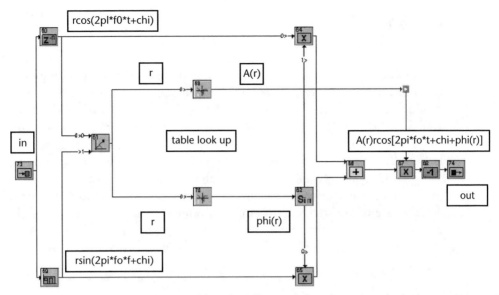

Figure 11.8   RF simulation of a TWT tube. The Hilbert transform is used to obtain the quadrature component needed to implement the phase shift portion. Note that we add a delay to the inphase signal to account for the group delay of the Hilbert filter.

Table 11.2   Harmonic Spur Chart of a Mini-Circuits Mixer

	0			1	16	16	31	52	33	28	68	49	63
RF harmoinic order	1	—	23	—	29	23	30	42	43	38	56	35	48
	2	>95	61	59	59	61	66	71	70	72	64	>67	67
	3	>94	>72	69	>74	63	>73	72	>71	>74	>74	>74	67
	4	>96	75	>73	>73	>74	>74	>74	>74	>74	>74	>73	>75
	5	>97	>74	>74	>73	>73	>72	75	>73	>72	>74	>74	>73
	6	>90	>68	>74	>74	73	>73	>73	>75	>73	>73	>74	>74
	7	>89	>67	>67	>74	>75	>75	74	>74	>74	>75	>73	>73
	8	>90	>67	>69	>67	>73	>74	>74	>73	>73	>73	>74	>74
	9	>89	>68	>68	>68	>67	>74	>74	>73	>74	>72	>74	>73
	10	>91	>68	68	>68	>68	>68	>74	>74	>74	>71	>74	>74
		0	1	2	3	4	5	6	7	8	9	10

$$x = A\sin(2\pi f_1 t) + A\sin(2\pi f_2 t)$$

Then using standard trigonometric identities reduce all terms from a power to a multiple angle as we did for the simple cubic model. Finally, gather up all terms with the same frequency. The coefficient of such a term will contain several of the model coefficients a_k. Then equate each such term to the corresponding value in the data table. The final result is a system of 11 equations in 11 unknowns that can be inverted. These coefficients can then be stored once and for all without need for recalculation each time.

11.4   Conclusion

This chapter was devoted to the simulation of nonlinear amplifiers. We started with the definitions of the standard impairments of the 1-dB compression point and the two-tone third order intercept point. These numbers are commonly presented in the amplifier specification sheets. We described a simple model using a third order polynomial and derived the coefficients to meet the amplifier specifications. Another amplifier model described was the TWT amplifier used in satellite-based transponders. Finally, we briefly described the methodology for obtaining another performance metric, namely the mixer spur chart. The basic idea, know as harmonic balance, represents the mixer as a high order polynomial. Into this polynomial a two signal is entered. By algebra, and the trigonometric identities, one can gather up all harmonic terms with their coefficients being a function of the polynomial coefficients and the input signal amplitudes. This results in a matrix equation that can be easily solved.

Selected Bibliography

Saleh, A. A. M., "Frquency-Independent Nonlinear Models of TWT Amplifiers," *IEEE Trans. On Comm.*, Vol. COM-29, No. 11, November 1981, pp. 1725–1720.

CHAPTER 12

Baseband Simulation

Here is a common simulation scenario. Consider an IS-136 wireless system. The carrier frequency is in the 880-MHz range, while the symbol rate is 24.3 Ksps. To implement such a system, the sample rate must be at least four times the carrier frequency, or 3.52 Gsps. But this is a huge overkill with respect to the information rate. In this chapter we develop the concept of baseband simulation, which eliminates the carrier altogether. We sample at a rate that is commensurate with the information bandwidth, thus greatly reducing the simulation time. Note, however, that this tactic works only if there are no nonlinear elements in the system such as amplifiers with IP_3, or when evaluating the effects of a frequency fade channel. In those cases the results are dependant on the actual carrier frequency, and we are stuck with the complete sampling system.

12.1 Basic Concept

We start with any system where the information is in the baseband forms $I(t)$ and $Q(t)$. We modulate this signal on a carrier to obtain the transmitted signal

$$s(t) = I(t)\cos(2\pi f_0 t + \theta) + Q(t)\sin(2\pi f_0 t + \theta)$$

where θ represents an unknown phase of the received carrier with respect to the carrier used in the I/Q down conversion.

At the receiver, we down convert $s(t)$ back to baseband using the quadrature technique as described in Chapter 6. After filtering out the $2f_0$ term, we have the estimates of the symbols $I(t)$ and $Q(t)$

$$I'(t) = I(t)\cos\theta + Q(t)\sin\theta$$
$$Q'(t) = I(t)\sin\theta - Q(t)\sin\theta$$

The whole trick now is to represent I and Q as complex numbers

$$z(t) = I(t) + jQ(t)$$
$$z'(t) = I'(t) + jQ'(t)$$

Now we can also write the phase angle term as $e = \cos\theta + j\sin\theta$. From all of this we get the neat result

163

$$z'(t) = z(t)e^{j\theta}$$

In other words an overall carrier phase shift is accomplished by a complex phase rotation. Furthermore, the rotation angle θ need not be constant. For one thing, what the transmitter thinks is the frequency f_0 is not necessarily what the receiver thinks is f_0. No two oscillators have exactly the same frequency. Furthermore, there may be Doppler frequency shift due to relative platform motion. In the signal intercept world, the receiver sees a "blob" of energy at some frequency and can only estimate the carrier. In all cases one can write the phase θ in the form

$$\theta(t) = \theta_0 + 2\pi\Delta ft$$

with Δf representing this unknown frequency difference.

The physical signal $s(t)$ is obtained by the relation

$$s(t) = \text{Re}\big[z(t)\big]$$

How this complex arithmetic is handled depends on the basic simulation engine. In some cases the compiler can handle complex numbers as a single entity, and the simulation would look like Figure 12.1.

In other simulation engines, a complex number is not recognized as a single unit. In this case, the real and imaginary parts must be kept separate with the algorithm performing the appropriate calculations, as shown in Figure 12.2.

12.2 Pass-Band Filtering

Returning to the original carrier-based simulation; there is always some form of pass-band filter at the carrier in the system. How do we handle this in our baseband simulation? It turns out that any pass-band filter impulse response can be written in the form

$$b_{pb}(t) = M(t)\cos(2\pi f_0 t) + N(t)\sin(2\pi f_0 t)$$

Figure 12.1   Simulation using complex numbers as a single entity.

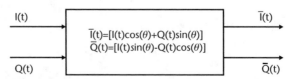

Figure 12.2   Complex arithmetic with the real and imaginary components kept separate.

where $M(t)$ and $N(t)$ are the low-pass components of the filter, which depend on the filter BW, type, and number of poles, but not the carrier. With this representation, let us see how a pass-band signal is transformed through this filter:

$$s'(t) = h_{pb}(t) * s(t)$$
$$= \left[M(t)\cos(2\pi f_0 t) + N(t)\sin\left(2\pi \bar{f}_0 t\right) \right] * \left[I(t)\cos(2\pi f_0 t) + Q(t)\sin(2\pi f_0 t) \right]$$
$$= \int \left[M(t-\tau)\cos\{2\pi f_0(t-\tau)\} + N(t-\tau)\sin\{2\pi f_0(t-\tau)\} \right]$$
$$\left[I(\tau)\cos(2\pi f_0 \tau) + Q(\tau)\sin(2\pi f_0 \tau) \right] d\tau$$

The next few steps of the development are straightforward (a classic cliché in text books). The final result is

$$s'(t) = \left\{ \int \left[M(\tau)I(t-\tau) + N(\tau)Q(t-\tau) \right] d\tau \right\} \cos(2\pi f_0 t)$$
$$+ \left\{ \int \left[M(\tau)Q(t-\tau) - N(\tau)I(t-\tau) \right] d\tau \right\} \sin(2\pi f_0 t)$$

In short-hand notation we can write

$$I'(t) = M(t) * I(t) + N(t) * Q(t)$$
$$Q'(t) = M(t) * Q(t) - N(t) * I(t)$$

From this relation the baseband equivalent filtering operation is shown in Figure 12.3.

In many cases the pass-band filter has symmetry such that $M(t) = 0$, which simplifies the operation further. Finally we note that the pass-band filters are usually set wide with respect to the modulation bandwidth. Thus, the effect of such a filter is not an issue; the operation can be done away with in it entirely.

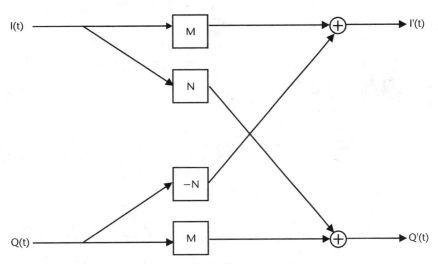

Figure 12.3 Baseband equivalent of a bandpass filtering operation.

12.3   Baseband Noise

We have shown how to handle the baseband equivalent of the signal portion of the simulation, but what about noise? It turns out that the pass-band noise can also be written in the common form

$$n(t) = n_I(t)\cos(2\pi f_0 t) + n_Q(t)\sin(2\pi f_0 t)$$

where the pass-band noise components have the statistics

$$\overline{n_I(t)} = \overline{n_Q(t)} = \overline{n_I(t)n_Q(t)} = 0$$
$$\overline{n_I^2(t)} = \overline{n_Q^2(t)} = N_0 B$$

and B is the bandwidth of the pass-band. One nice thing to remember is that a complex constant phase rotation on the noise produces another set of noise with exactly the same statistics. This is due to the fact that any linear combination of a Gaussian-distributed signal is again Gaussian-distributed (this is a good exercise for the interested reader).

In Figure 12.4 we show a baseband simulation of a QPSK signal with RRC baseband (not pass-band) filtering.

The noise variance is determined in a manner similar to that described in Chapter 9. For a specific E_b/N_0, we obtain the result for each of the noise components:

$$N_0 = \overline{\left[I^2(t) + Q^2(t)\right]}\Big/\left[E_b/N_0\right]R$$

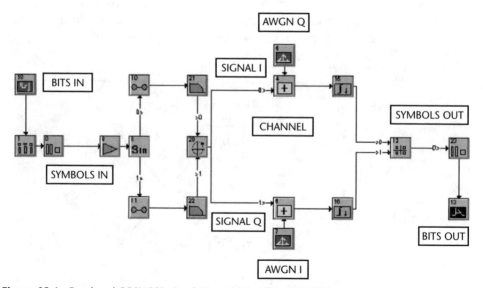

Figure 12.4   Baseband QPSK BER simulation with baseband RRC filters.

12.4 Conclusion

Baseband simulations are desirable to reduce the simulation time. Unless the system contains a nonlinear element of some sort, such a simulation is possible. In this chapter we developed the concept of a baseband simulation in terms of the I and Q components of the signal. It was shown that a complex representation of the signal $z = I + jQ$ is very useful in developing the required processing steps.

Selected Bibliography

Pahlavan, K., and A. H. Leveseque, *Wireless Information Networks*, New York: John Wiley & Sons, 1995.

Proakis, J. B., *Digital Communications*, New York: McGraw-Hill 1983.

Rappaport, T. S., *Wireless Communications, Principles and Practice*, Upper Saddle River, NJ: Prentice Hall, 2002.

Sklar, B., *Digital Communications, Fundamentals and Applications*, Upper Saddle River, NJ: Prentice Hall, 2001.

Steele, R., *Mobile Radio Communications*, London: Pentech Press Limited, 1992.

CHAPTER 13

Ultra-Wideband Systems

UWB systems are currently undergoing intense development by a large number of commercial organizations worldwide. The goal is to provide a very high data rate for applications in wireless communications. In this chapter we present the current state of development in the field. The system concepts are still under a state of flux and what finally emerges may differ from what is presented here. However, most of the basic concepts should not change.

The first observation is the nature of wireless use and capability. From Chapter 5 on detection theory we noted that the output SNR of the detection matched filter is

$$SNR = 2E_b/N_0 = A^2/N_0R$$

for a constant envelope signal of amplitude A, and data rate R. Clearly, as R increases, the SNR goes down for a fixed A. To make up for the loss, A must increase accordingly, but A^2 is the signal power that cannot be increased without limit. The bottom line is that the wireless UWB applications are intended for a very short range. The terminology is called wireless personal area networks (WPANs), which operate over distances of the order of room size.

What is a UWB system? The following is the current definition:

- Bandwidth contained within 3.1 to 10.6 GHz (communication sys.);
- Maximum EIRP of −41.3 dBm/MHz;
- UWB BW defined by

$$2\frac{f_u - f_1}{f_u + f_1} \geq 2$$

or BW ≥ 500 MHz;

- No modulation scheme is implied.

Figure 13.1 shows the two-band allocation for UWB systems. The low band occupies 3.1 to 5.0 MHz, while the high band contains 5.0 to 9.8 MHz.

Currently there are two very different competing systems under development. We shall cover each in turn.

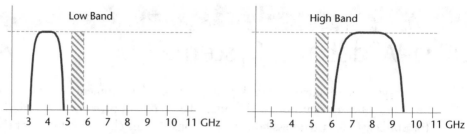

Figure 13.1   UWB spectral allocation: (a) low band; and (b) high band.

13.1 Direct Sequence DS-UWB Systems

The DS-UWB system is much like the DS spread spectrum concept discussed in Chapter 6. The data bits are replaced by code words, as shown in Section 13.1.5. Figure 13.2 is the general block diagram of the system. The remaining sections provide the pertinent details.

13.1.1   Scrambler

Most wireless systems use a scrambler. Note that this is not the same as an interleaver. The idea of the scrambler is to preclude transmitting long strings of [1] or [0] bits. Figure 13.3 shows this operation. The feedback system is just as described in Chapter 6 regarding DSSS systems. The polynomial in this case is

$$g(D) = 1 + D^{14} + D^{15}$$

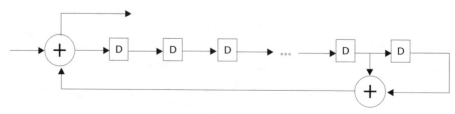

Figure 13.2   DS-UWB data block diagram.

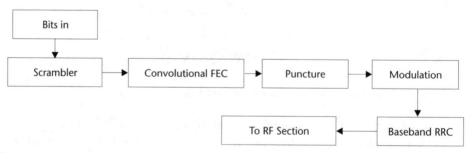

Figure 13.3   DS-UWB scrambler.

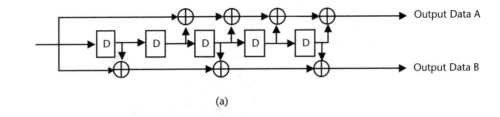

(a)

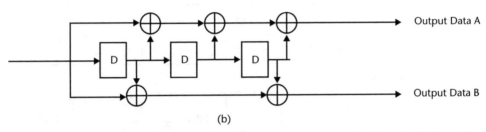

(b)

Figure 13.4 (a) Rate 1/2, $K = 6$ convolutional encoder code polynomial = $(65,57)_8$. (b) Rate 1/2, $K = 4$ convolutional encoder code polynomial = $(15,17)_8$.

The only difference here is that the input data [b] is running at the same rate as the shift back register. In other words, there is no expansion of bandwidth here. The output of the scrambler is given by

$$S_0 = x_0 \otimes b_0$$

13.1.2 Forward Error Correction

The DS-UWB employs two different convolutional codes as shown in Figure 13.4(a, b).

13.1.3 Puncturing

Both of the encoders are rate 1/2. This means that there are two output bits for every one input bit, which doubles the required transmission bandwidth. In many cases this is unacceptable, and a lower rate such as 2/3 or 3/4 is desired. One way to do this is to directly implement codes with the desired rates. While this is possible, it is almost never done. Strangely enough, what is done is to take the rate 1/2 or 1/3 code, and simply throw away bits. This deletion is not done randomly but at places where the simulations and theory show that the net effect is negligible. Figure 13.5 shows the operation that converts $r = 1/2$ to $r = 3/4$, which is required for some UWB data modes. The encoder accepts 9 input bits into the $r = 1/2$ convolutional encoder, producing 18 output bits. Six of these 18 bits are eliminated, sending 12 bits into the channel. Nine bits in, 12 bits out or $r = 3/4$.

The reinsertion of the dummy bit is a logical zero with the possible data being [1, −1]. The Viterbi decoder works on soft decisions in this operation. It is also possible to modify the FEC coder to ignore the trellis coder paths and not reinsert the data bits (see Appendix C on error correcting codes).

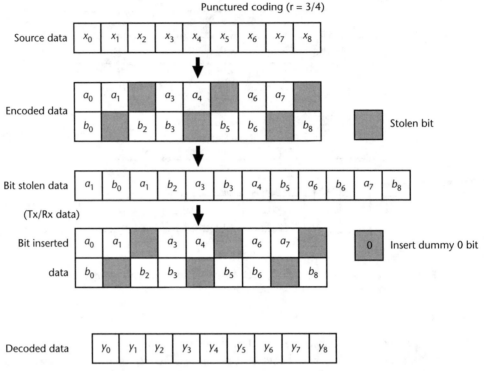

Figure 13.5 Rate 1/2 to 3/4 puncturing operation.

13.1.4 Interleaver

The interleaver employed is called a convolutional interleaver. It works differently than the simple row x column algorithm mentioned in Chapter 1. Figure 13.6 is a block diagram.

The block interleaver mentioned in Chapter 1 works on groups of bits or blocks at a time. By contrast, the convolutional interleaver works in a time continuous mode. There is a series of N registers of depth 0 (direct feed through), J, $2J$, ..., $(N-1)J$. The two commutator switches cycle switch in unison from one register to the next on a bit-by-bit basis. The registers are initialized to 0 at time $t = 0$.

At time $t = 0$, the first bit is transmitted through. At $t = 1$, the commutators drop to the second register and read out the rightmost bit, which is a 0 (and will eventually be ignored in the deinterleaver). At the same time the second bit is entered into the leftmost position of the register. At time $t = 2$, the switches drop to the third posi-

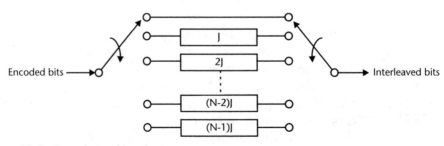

Figure 13.6 Convolutional interleaver.

tion and read out the rightmost bit, which again is a 0. Bit 2 is entered into the leftmost position of this register. This process continues to the Nth register. The commutators now go back up to the top and the whole process is repeated (i.e., the N+1 bit is transmitted). As the process continues, the bits in the various registers get pushed to the right, and are eventually read out into the system.

The advantage of the convolutional interleaver is that the throughput delay and memory cells required are less than the corresponding block interleaver.

The UWB interleaver uses $J = 7$, $N = 10$.

13.1.5 Modulation

The DS-UWB system employs a combination of modulation and coding (not FEC here). The primary form of modulation is called BPSK. But this is somewhat of a misnomer, as we shall see shortly. The optional form is called 4-biorthogonal keying (4-BOK), which requires some explanation

4M-BOK Coding
The basic idea of orthogonal keying (OK) is to take a group of k information bits $[m_k]$ and produce a 2^k bit code word $[R_{n,k}]$ such that all code words are orthogonal; that is,

$$\sum_{k=0}^{N} R_{nk} {}^* R_{mk} = 0 \quad n \neq m$$

$$= 1 \quad n = m$$

One procedure for generating such codes is simple and based on the Walsh Haddamard transformation. We start with $k = 1$,

$$[0] \rightarrow [1,0]$$
$$[1] \rightarrow [0,1]$$

Now we define the Walsh matrix as

$$[W_2] = \begin{bmatrix} 1 & 0 \\ 0 & 1 \end{bmatrix}$$

Using Boolean logic, the two rows of this matrix are seen to be orthogonal. From here we generalize as follows:

$$[W_{2k}] = \begin{bmatrix} [W_k], & [W_k] \\ [W_k], & \overline{[W_k]} \end{bmatrix}$$

and so on. The IS-95 wireless CDMA system uses such codes to provide strict orthogonality between multiple users who coexist in the same bandwidth.

In BOK coding, start with the OK codes just described but augment the code set $[W_k]$ by adding the inverse of each of the original orthogonal codes. This is called biorthogonal coding in most textbooks. The second set of code words is the same as the first set but with the leading bit set to 1. Specifically, 8-BOK makes the assignment as shown in Table 13.1.

The DS-UWB system actually uses 4-BOK, which has the coding shown in Table 13.2.

This is a strange looking table since both columns are the same. However, the right-hand column is only an example and there are other choices.

We can restate Table 13.2 in the more general form of Table 13.3.

The only requirements on $S_1(t)$ and $S_2(t)$ are

$$\int_0^T S_1(t)S_2(t)dt = 0$$

$$\int_0^T S_1^2(t)dt = \int_0^T S_2^2(t)dt = \text{Energy}$$

The DS-UWB uses *ternary* codes, which have three states [+, 0, –]. The 0 implies no transmission during that chip time. The codes for the various data rates and modulation are shown in the Tables 13.4 through 13.7 below.

The term BPSK now implies only 1 input bit per output code sequence L. The signal is not strictly BPSK since there are three levels. A more appropriate term is *antipodal*, which requires $S_2(t) = -S_1(t)$.

In 4-BOK, 2 input bits are encoded into a cover sequence L. Over the air the modulation is the same as normal BPSK except the code bit 0 simply transmits a blank or no signal.

To see how all of this plays out, consider the 28-Mbps (nominal data rate) case in the lower band, Table 13.4. This rate uses an $L = 24$ code. Now we wish to use piconet channel 1 center frequency of 3,939 MHz (Table 13.8), which requires spreading code 1 (Table 13.9), specifically [–1, 0, 1, –1, -1, -1, 1, 1, 0, 1, 1, 1, 1, -1, 1, -1, 1, 1, 1, -1, 1, -1, -1, 1].

We now have the following transmission chip rate calculation:

Chip rate = [28 Mbps] × [2 (for FEC $r = 1/2$)] × [24 (L code)] = 1,344 Mbps

Table 13.1 8-BOK Binary Coding

Input Bits	Coded Bits
000	0000
001	0101
010	0011
011	0110
100	1111
101	1010
110	1100
111	1001

Table 13.2 4-BOK
Binary Coding

Input Bits	Coded Bits
00	00
01	01
10	11
11	10

Table 13.3 General
4-BOK Coding

Input Bits	Coded Signal
00	$S_1(t)$
01	$S_2(t)$
10	$-S_2(t)$
11	$-S_1(t)$

Table 13.4 Available Data Rates in Lower Operating Band for BPSK with Associated FEC and Spreading Code Rate

Nominal Data Rate (Mbps)	FEC Rate	Code Length (L)	Bits per Symbol	Symbol Rate
28	½	24	1	$F_{chip}/24$
55	½	12	1	$F_{chip}/12$
110	½	6	1	$F_{chip}/6$
220	½	3	1	$F_{chip}/3$
500	¾	2	1	$F_{chip}/2$
660	1	2	1	$F_{chip}/2$
1,000	¾	2	1	F_{chip}

Table 13.5 Available Data Rates in Lower Operating Band for 4-BOK with Associated FEC and Spreading Code Rates

Nominal Data Rate (Mbps)	FEC Rate	Code Length (L)	Bits per Symbol	Symbol Rate
110	½	12	2	$F_{chip}/12$
220	½	6	2	$F_{chip}/6$
500	¾	4	2	$F_{chip}/4$
660	1	4	2	$F_{chip}/4$
1,000	3/4	4	2	$F_{chip}/2$
1,320	1	2	2	$F_{chip}/2$

Note in Table 13.4 the chip rate is 3,939 MHz/3 = 1,313 MHz. We see that there is a disconnect between the two chip rates. The proper calculation is to start with Table 13.8 and work backward to find the associated data rate—27.354 Mbps

Table 13.6 Available Data Rates in Higher Operating Band for BPSK with Associated FEC and Spreading Code Rate

Nominal Data Rate (Mbps)	FEC Rate	Code Length (L)	Bits per Symbol	Symbol Rate
55	1/2	24	1	$F_{chip}/24$
110	1/2	12	1	$F_{chip}/12$
220	1/2	6	1	$F_{chip}/6$
500	3/4	4	1	$F_{chip}/4$
660	1	4	1	$F_{chip}/4$
1,000	3/4	4	1	$F_{chip}/2$
1,320	1	2	1	$F_{chip}/2$

Table 13.7 Available Data Rates in Higher Operating Band for 4-BOK with Associated FEC and Spreading Code Rate

Nominal Data Rate (Mbps)	FEC Rate	Code Length (L)	Bits per Symbol	Symbol Rate
220	1/2	12	2	$F_{chip}/12$
660	3/4	6	2	$F_{chip}/6$
1,000	3/4	4	2	$F_{chip}/4$
1,320	1	4	2	$F_{chip}/4$

Table 13.8 Piconet Channel Numbers with Associated Chip Rates and Carrier Frequency

Piconet Channel	Chip Rate (MHz)	Center Frequency (MHz)	Spreading Code Set
1	1,313	3,939	1
2	1,326	3,978	2
3	1,339	4,017	3
4	1,352	4,056	4
5	1,300	3,900	5
6	1,365	4,094	6
7	2,626	7,878	1
8	2,652	7,956	2
9	2,678	8,034	3
10	2,704	8,112	4
11	2,600	7,800	5
12	2,730	8,190	6

in this case. If we had chosen the carrier frequency of 3,978 MHz, then by the same logic the data rate would be 27.625 Mbps. This is why Tables 13.3 through 13.7 use the term "*Nominal* Data Rate." This terminology avoids the cumbersome book-keeping that would be necessary to detail all cases. All of these different actual rates only vary by a few percent.

Table 13.9 $L = 12$ and 24 Codes for BPSK and Acquisition

Code Set Number	$L = 24$	$L = 12$
1	-1. 0, 1, -1, -1, -1, 1, 1, 0, 1, 1, 1, 1, -1, 1, -1, 1, 1, 1, -1, 1, -1, -1, 1	0, -1 -1, -1, 1, 1, 1, -1, 1, 1, -1, 1
2	-1, -1, -1, -1, 1, -1, 1, -1, 1, -1, -1, 1, -1, 1, 1, -1, -1, 1, 1, 0, -1, 0, 1, 1	-1, 1, -1, -1, 1, -1, -1, -1, 1, 1, 1, 0
3	-1, 1, -1, -1, 1, -1, -1, 1, -1, 0, -1, 0, -1, -1, 1, 1, 1, -1, 1, 1, 1, -1, -1, -1	0, -1, 1, -1, -1 1, -1, -1, -1, 1, 1, 1
4	0, -1, -1, -1, -1, -1, -1, 1, 1, 0, -1, 1, 1, -1, 1, -1, -1, 1, 1, -1, 1,1 -1, 1, -1	-1, -1, -1, 1, 1, 1, -1, 1, 1, -1, 1, 0
5	-1, 1, -1, 1, 1, -1, 1, 0, 1, 1, 1, -1, -1, 1, 1, -1, 1,1, 1, -1, -1, -1, 0, -1	-1, -1, -1, 1, 1, 1, -1, 1, 1, -1, 1, 0
6	0, -1, -1, 0, 1, -1, -1, 1, -1, -1, 1, 1, 1, 1, -1, -1, 1, -1, 1, -1, 1, 1, 1, 1	0, -1, -1, -1, 1, 1, 1, -1, 1, 1, -1, 1

A similar procedure is used for the 4-BOK coding. For example, from Table 13.7 we desire a 220-Mbps nominal data rate for the upper band. This rate requires an FEC rate $r = 1/2$, with an $L = 12$ code set. The chip rate calculation is then

Chip rate = [220 Mbps] × [2 FEC] × [2 bits/symbol] × [$L = 12$] = 2,640 Mbps

From Table 13.8, we select upper band piconet 7, using a frequency of 7,878 MHz. The corresponding chip rate is then 2,626 Mbps. As before, there is a mismatch. The actual data rate would be 220 × (2,626/2,640) = 218.83 Mbps. Table 13.10 describes at $L = 6$.

For each 2 bits of input data, the particular $L = 12$ spreading code is shown in Table 13.11. The reader can verify that this code set has the required 4-BOK properties.

The final time domain baseband waveform for any case can be written in the form

$$s(t) = \sum_{k=0}^{L-1} a_k b_{rrc}(t - kT_c)$$

where $a_k = [1, 0, -1]$ according to the to the entry tables below.

Note the chip rate = carrier frequency/3.

Figure 13.7 is a single block diagram of the modulation process. The two bit data group selects one of the four cover codes for L = 24 as shown in Figure 13.8 for the two bit pattern [0, 0]. Each modulated bit (impulse) drives the RRC Filter in a manner described in Chapter 8. The carrier modulated signal is shown in Figure 13.9. Finally the signal PSD is shown in Figure 13.10.

Table 13.10 $L = 6$ and Shorter Codes for BPSK

Code Set Numbers	$L = 6$	$L = 4$	$L = 3$	$L = 2$	$L = 1$
1 – 6	1, 0, 0, 0, 0, 0	1, 0, 0, 0	1, 0, 0	1, 0	1

Table 13.11 *L* = 12 and Shorter Codes for 4-BOK Code Sets 1 to 6

Input Data Gray Coding	*L* = 12 BOK	*L* = 6 BOK	*L* = 4 BOK	*L* = 2 BOK
00	1,0, 0, 0, 0, 0, 0, 0, 0, 0, 0, 0	1, 0, 0, 0, 0, 0	1, 0, 0, 0	1, 0
01	0,0, 0, 0, 0, 0, 0, 1, 0, 0, 0, 0	0, 0, 0, 1, 0, 0	0, 0, 1, 0	0, 1
11	−1,0, 0, 0, 0, 0, 0, 0, 0, 0, 0, 0	−1, 0, 0, 0, 0, 0	−1, 0, 0, 0	−1, 0
10	0,0, 0, 0, 0, 0, 0, −1, 0, 0, 0, 0	0, 0, 0, −1, 0, 0	0, 0, −1, 0	0, −1

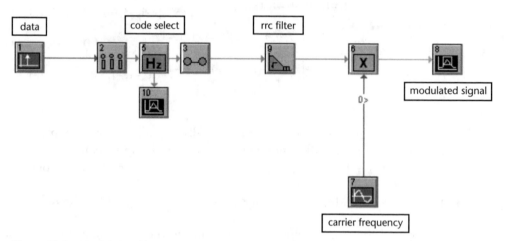

Figure 13.7 Modulation block diagram.

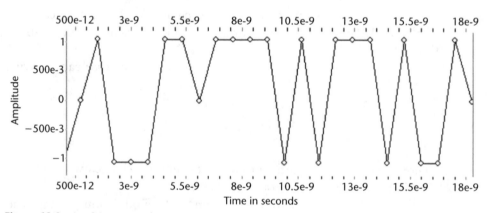

Figure 13.8 *L* = 24 cover code.

13.2 Multiband OFDM UWB-Multiband OFDM Alliance Systems

The alternate UWB concept to DS is based on orthogonal frequency division modulation (OFDM) modulation as discussed in Chapter 6. For the purpose of operation, the UWB spectra of Figure 13.1 is broken into five band groups. The first four groups have three subbands, and the fourth has two subbands. Figure 13.11 shows this assignment. The use of these bands is discussed in the section on RF modulation

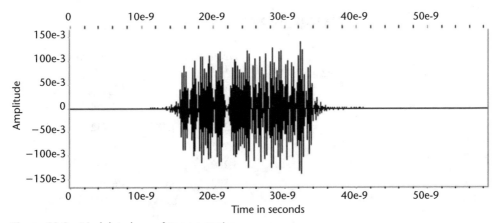

Figure 13.9 Modulated waveform on carrier.

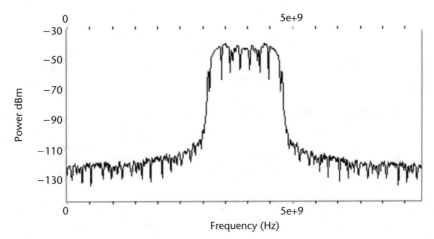

Figure 13.10 PSD of modulated signal.

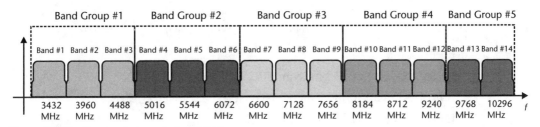

Figure 13.11 Band assignments for OFDM.

below (Section 13.2.3). Table 13.12 gives the specific frequency assignments for each.

The details of the first line in Table 13.13 are as follows:

- Input data rate: 160/3 = 53.3 Mbs;

Table 13.12 OFDM Frequency Band Assignments

Band Group	BAND_ID	Lower Frequency (MHz)	Center Frequency (MHz)	Upper Frequency (MHz)
1	1	3,168	3,432	3,696
	2	3,696	3,960	4,224
	3	4,224	4,488	4,752
2	4	4,752	5,016	5,280
	5	5,280	5,544	5,808
	6	5,808	6,072	6,336
3	7	6,336	6,600	6,864
	8	6,864	7,128	7,392
	9	7,392	7,656	7,920
4	10	7,920	8,184	8,448
	11	8,448	8,712	8,976
	12	8,976	9,240	9,504
5	13	9,504	9,768	10,032
	14	10,032	10,296	10,560

Table 13.13 OFDM Modulation Parameters

Data Rate (Mbps)	Modulation	Coding Rate (R)	Conjugate Symmetric Input to IFFT	Time Spreading Factor (TSF)	Overall Spreading Gain	Coded Bits per OFDM Symbol (N_{CBPS})
53.3	QPSK	1/3	Yes	2	4	100
80	QPSK	1/2	Yes	2	4	100
110	QPSK	11/32	No	2	2	200
160	QPSK	1/2	No	2	2	200
200	QPSK	5/8	No	2	2	200
320	QPSK	1/2	No	1 (No spreading)	1	200
400	QPSK	5/8	No	1 (No spreading)	1	200
480	QPSK	3/4	No	1 (No spreading)	1	200

- QPSK modulation, 2 bits/sym: 80/3 = Msym/sec;
- Rate 1/3 FEC: [80/3]*3 = 80 Msym/sec;
- OFDM time: 0.3125 usec;
- Required data symbols per OFDM frame: 0.3125 usec*80 Msym/sec = 50;
- Conjugate addition: yes;
- Available data symbols per OFDM frame: 2*25 = 50.

13.2.1 FEC Coding

Figure 13.12 shows the basic $K = 7$, rate = 1/3 Viterbi encoder used.

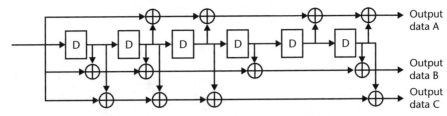

Figure 13.12   FEC coding with code polynomials $[133]_8$, $[165]_8$, $[171]_8$.

13.2.2 Puncturing

Figures 13.13 through 13.16 show the details of the puncturing of the basic rate 1/3 code for the various output data rates.

13.2.3 Modulation

Once the data is finally encoded and interleaved, it is ready for modulation. The modulation format is OFDM, as discussed in Chapter 6. In this case the system is based on an $N = 128$ point FFT. An OFDM symbol is written in the form

$$s(t) = \sum_{k=0}^{127} c_k e^{2\pi jk\Delta_f t}$$

where $\Delta f = 528$ MHz/128 = 4.125 MHz. The time extent of this IFFT is the reciprocal of the spacing, or 242.42 ns. The total OFDM frame time is

165/528 MHz = 312.5 ns. Unlike the previous discussion on OFDM in Chapter 6, there is no periodic extension of the tones to fill the gap. In this case the signal is

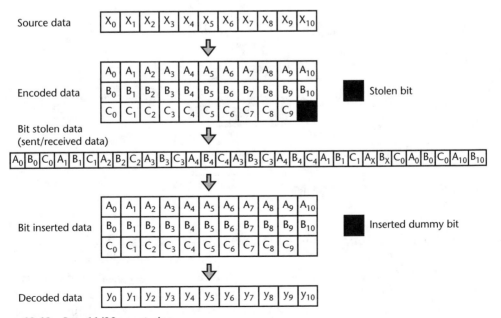

Figure 13.13   Rate 11/32 puncturing.

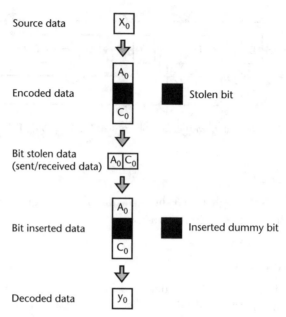

Figure 13.14   Rate 1/2 puncturing.

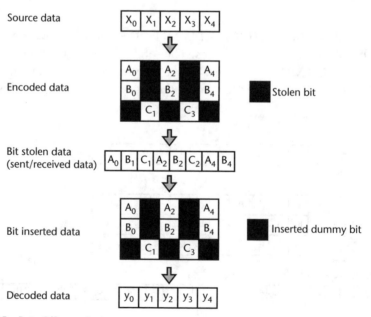

Figure 13.15   Rate 5/8 puncturing.

simply blanked out over the remaining $(165 - 128)/528$ MHz $= 37/528$ MHz $= 70.08$ ns. Table 13.14 shows the various parameters under discussion.

Out of the 128 possible tones, only 100 subcarriers are used for data, with each being modulated by 2 complex bits or QPSK. Twelve more subcarriers are devoted to pilot signals used for synchronization. These pilot tones are defined for the kth symbol as follows:

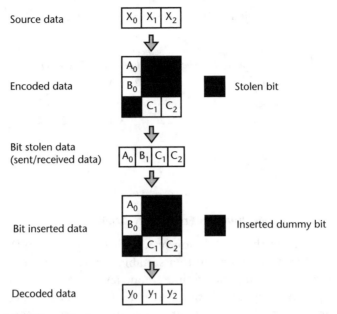

Figure 13.16   Rate 3/4 puncturing.

Table 13.14   OFDM Modulation Parameters

Parameter	Value
N_{SD}: Number of data subcarriers	100
N_{SDP}: Number of defined pilot carriers	12
N_{SG}: Number of guard carriers	10
N_{ST}: Number of total subcarriers used	122 $(= N_{SD} + N_{SDP} + N_{SG})$
Δ_F: Subcarrier frequency spacing	4.125 MHz $(= 528\text{ MHz}/128)$
T_{FFT}: IFFF/FFT period	242.42 ns $(1/\Delta_F)$
T_{ZP}: Zero pad duration	70.08 ns $(=37/528\text{ MHz})$

$$P_{n,k} = p_k \left(1+j\right)/\sqrt{2}\, n = 25,55$$
$$(-1-j)/\sqrt{2}\, n= 5,15,35,45]$$

where p_k is a member of an 127 bit pseudo-random bit sequence. This sequence is repeated modulo 127. This defines 6 of the 12 subcarriers. The remaining six are given by

$$P_{-n,k} = P_{n,k}^* \quad \text{for bit rate } < 106.7\,Mbps$$
$$P_{-n,k} = P_{n,k} \quad \text{for bit rate } > 106.7\,Mbps$$

There are now $[128 - (100 + 12)] = 16$ subcarriers still unoccupied. Ten of these (five on each end) are called guard subcarriers.

Table 13.15   TFC Frequency Code Parameters

TFC Number	Preamble Pattern Number	Cover Sequence Number	Length 6 Time Frequency Code (BAND_ID Values for Band Group 1)					
1	1	1	1	2	3	1	2	3
2	2	1	1	3	2	1	3	2
3	3	2	1	1	2	2	3	3
4	4	2	1	1	2	2	3	3
5	1	2	1	2	1	2	1	2
6	2	2	1	1	1	2	2	2

13.2.4   Carrier Modulation Frequency Agility

The final step in the modulation process is to take the OFDM symbols as defined above and translate them up to a suitable carrier frequency. To increase capacity, the system uses what is called time frequency codes (TFC). This is essentially a simple frequency hopping scheme. Table 13.15 shows the specifics. For each of the band groups 1 to 4, there are three subbands. There are four hopping patterns, as shown in the first four entries of the table. So if band 1 is being employed with TFC 2, then the carrier frequencies would be 3,166, 4,224, 3,696, 3,166, 4,224, and 3,696 MHz. The pattern repeats itself in groups of six. For band group 5, there are two hopping patterns: TFC 5 and 6 in the table. In total for each band group there are $4 \times 4 + 2 \times 2 = 18$ separate piconets that can be assigned.

Many military and wireless systems use some form of orthogonal frequency hopping to reduce mutual interference between the various users.

13.3   Conclusion

This chapter dealt with the new wireless technology known as ultra-wideband. There are two competing modulation technologies. The first is based on direct sequence techniques using the BOK coding concept. The second is based on OFDM technologies currently used in 802.11g WLANs.

Selected Bibliography

DS-UWB Physical Layer Submission to IEEE 802.15 Task Group 3a, IEEE P802.15-04/ 0137r00r00137r0, Wireless Personal Area Networks, March 2004.

DS-UWB Physical Layer Submission to IEEE 802.15 Task Group 3a, IEEE P802.15-04/ 0137r3, Wireless Personal Area Networks, July 2004.

Multi-band OFDM Physical Layer Proposal for IEEE 802.15 Task Group 3a, IEEE P802.15-03/268r0, Wireless Personal Area Networks, July 2003.

Table of Laplace Transforms

Table A.1 Laplace Transforms

Time Function	Laplace Function
$f(t)$	$F(s) = \int_0^\infty f(t)e^{-st}dt$
$f(t) = \int_{c-j\infty}^{c+j\infty} F(s)e^{st}ds$	$F(s)$
$Af_1(t) + Bf_2(t)$	$Af_1(s) + BF_2(s)$
$\int_0^t f_1(t-\tau)f_2(\tau)d\tau$	$F_1(s)F_2(s)$
$Df(t)/dt$	$sF(s) - f'(0)$
$\int_0^t f(t)dt$	$F(s)/s$
$-tf(t)$	$dF(s)/ds$
$f(t)/t$	$\int_s^\infty F(s)ds$
$e^{at}f(t)$	$F(s-a)$
$u(t)$	$1/s$
t	$1/s^2$
e^{-at}	$1/(s+a)$
$\sin\omega t$	$\omega/[s^2 + \omega^2]$
$\cos\omega t$	$s/[s^2 + \omega^2]$

Elements of Probability Theory and Random Variables

B.1 Probability Theory

Probability theory is what is known as an axiomatic system. The basic idea is to define a collection of objects or sample space with each element being a sample point. A set of rules (axioms) regarding these objects is then postulated. From there, the system is expanded with additional theorems and properties. The nature of the objects and the axioms are totally at the discretion of the developer. Hopefully a useful system results.

B.1.1 Axioms of Probability

The axioms of probability are as follows:

- Axiom 1: Given an experiment, there exists a sample space $\{S\}$ that represents all possible outcomes of an experiment, and a subset $\{A\}$ of $\{S\}$ called events.
- Axiom 2: For each event $\{A\}$ there is an assigned probability of that event, such that $P\{A\} = 0$.
- Axiom 3: The probability of the whole space is $P\{S\} = 0$.
- Axiom 4: If two events $\{A\}$ and $\{B\}$ are mutually exclusive, then

$$P\{A\} \bigcap P\{B\} = \{\text{null set}\}$$
$$P\{A \bigcup B\} = P\{A\} + P\{B\}$$

If $\{A\}$ and $\{B\}$ are not mutually exclusive, then the following holds:

$$P\{A \bigcup B\} = P\{A\} + P\{B\} - P\{A \bigcap B\}$$

B.1.2 Dice Example

As an example of such a sample space, consider the game of chance, which involves two dice, each with sides numbered from 1 to 6. Table B.1 is one possible sample space that involves 6*6 = 36 objects. In this sample space we can ask questions (experiment) regarding both dice individually (e.g., what is the probability of the event that die A is even (2, 4, 6) and die B is greater than 3). In Table B.1 this would be events [10, 11, 12, 22, 23, 24, 34, 35, 36].

In the casino game of craps, the only issue is what the sum is of the spots on the dice as noted in the last column of Table B.1. In this case we can derive a second

Table B.1   Sample Space of Two Dice

Event #	Die A	Die B	Sum	Probability
1	1	1	2	1/36
2	1	2	3	1/36
3	1	3	4	1/36
4	1	4	5	1/36
5	1	5	6	1/36
6	1	6	7	1/36
7	2	1	3	1/36
8	2	2	4	1/36
9	2	3	5	1/36
10	2	4	6	1/36
11	2	5	7	1/36
12	2	6	8	1/36
13	3	1	4	1/36
14	3	2	5	1/36
15	3	3	6	1/36
16	3	4	7	1/36
17	3	5	8	1/36
18	3	6	9	1/36
19	4	1	5	1/36
20	4	2	6	1/36
21	4	3	7	1/36
22	4	4	8	1/36
23	4	5	9	1/36
24	4	6	10	1/36
25	5	1	6	1/36
26	5	2	7	1/36
27	5	3	8	1/36
28	5	4	9	1/36
29	5	5	10	1/36
30	5	6	11	1/36
31	6	1	7	1/36
32	6	2	8	1/36
33	6	3	9	1/36
34	6	4	10	1/36
35	6	5	11	1/36
36	6	6	12	1/36

Table B.2   Sample Space of the Sum of Two Dice

Sum	Events	Probability
2	1	1/36
3	2,7	2/36
4	3, 8, 13	3/36
5	4, 9, 14,19	4/36
6	5,10, 15, 20,25	5/36
7	6, 11, 16, 21, 26, 31	6/36
8	12, 17, 22, 27, 32	5/36
9	18, 23, 28, 33	4/36
10	24, 29, 34	3/36
11	30, 35	2/36
12	36	1/36

sample space as shown in Table B.2. Furthermore, we can define, for example, an event that has as its outcome the sum of the two spots on the dice. This is also shown in Table B.2

We can also reduce the sample space to 11 elements that represent the sum of the first column, simplifying further calculations. For example, when initiating a game of casino dice (craps), if the shooter roles a 7 or 11, he wins. What is the probability of that event? From Table B.2 we see that

$$P\{7 \cup 11\} = P\{7\} + P\{11\} = 6/36 + 2/36 = .222$$

where we have made use of Axiom 4. By the rules, if the shooter instead roles a {2, 3, 12}, he loses. The probability of that event is $1/36 + 2/36 + 1/36 = 0.111$.

B.1.3   Conditional Probability

A very common situation in probability problems is the following: given that an event {A} has occurred, what is the probability that a second event {B} will occur? The notation used is $P[\{B\}|\{A\}]$, and the basic definition is

$$P\big[\{B\}|\{A\}\big] = P\big[\{B\} \cap \{A\}\big]/P\big[\{A\}\big]$$

Now since $P[\{A\}I\{B\}] = P[\{B\}I\{A\}]$, we can write [Baye's Theorem]

$$P\big[\{B\}|\{A\}\big]P\big[\{A\}\big] = P\big[\{B\} \cap \{A\}\big] = P\big[\{B\} \cap \{A\}\big] = P\big[\{A\}|\{B\}\big]P\big[\{B\}\big]$$

This is a significant result. What it allows us to do is to switch which event is conditioned on the other. These terms are sometimes denoted as a priori and a poseteri. Recall that in Chapter 5 on detection theory we wanted to choose a signal.

Such that $P[A,1N] \leq P[A_2 1r]$ we used the above result to change the detection requirement $P[r,1A_1] \leq P[r1A_2]$. This probability could be calculated.

B.2   Random Variables

In the simplest sense, a random variable is any entity that can only be described in terms of some statistical quantities. The starting point is called the *probability distribution function* or the *cumulative distribution function, $F_x(\alpha)$*, which is defined by the probability of the outcome of an experiment that produces a value for the random variable x is less than α. Do not confuse the name of the random variable x with a value that it can have, α. From this definition we have the following restrictions on $F_x(\alpha)$:

$$0 \leq F_x(\alpha) \leq 1$$

and if $\alpha > \beta$,

$$F_x(\alpha) > F_x(\beta)$$

which says that we cannot have negative probability. It is more common to deal with the probability distribution function (PDF), $f_x(\alpha)$ defined as the derivative of $F_x(\alpha)$

$$f_x(\alpha) = dF_x(\alpha)/d\alpha$$

We interpret the relation $f_x(\alpha)d\alpha$ as the probability that the random variable x occurs from α to $\alpha + d\alpha$. Finally we must have

$$\int_{-\infty}^{\infty} f_x(\alpha)d\alpha = 1$$

Let $G(x)$ be a function of the random variable x. Then the expected value is given by

$$E[G(x)] = \int_{-\infty}^{\infty} G(\alpha)p_x(\alpha)d\alpha$$

This so far is a discussion of a single random variable. The idea can be extended to two random variables x and y through a cumulative distribution function $F_{x,y}(\alpha, \beta)$, which is the probability that the random variables x and y are less than α and β. Then the two-dimensional PDF is simply

$$p_{x,y}(\alpha,\beta) = \partial^2 F_{x,y}(\alpha,\beta)/\partial\alpha\partial\beta$$

Some useful general relations are

$$p_x(\alpha) = \int_{-\infty}^{\infty} p_{x,y}(\alpha,\beta)d\beta$$

and the conditional probability expression

$$p_x(\alpha|\beta = \varepsilon) = p_{x,y}(\alpha,\varepsilon)/p_y(\varepsilon)$$

B.2.1   The Gaussian Random Variable

The single most import random variable is the Gaussian PDF (GRV(μ,σ)) defined by

$$p_x(\alpha) = \exp\left[-(\alpha-\mu)^2/2\sigma^2\right]/\sqrt{2\pi\sigma^2}$$
$$\mu = E[x]$$
$$\sigma^2 = E\left[(x-\mu)^2\right]$$

This statement is based on the central limit theorem, which states that: If $[x_1, x_2, x_3, \ldots]$ are a sequence of independent random variables of any distribution, then the distribution of the sum

$$S_n = \sum_{k=1}^{n} x_k$$

converges to GRV(μ, σ) as $n \rightarrow \infty$.

A classic example of this involves the binomial distribution. Consider an experiment with two outputs: [1] with probability p, and [0] with probability $q = 1 - p$. Now perform this experiment n times and calculate the number of 1s that occur. To see how this works, consider the case $n = 3$. Then there are eight possible outcomes of the experiment, as shown in Table B.3.

From this table we get the results

$$P[0] = q^3, \quad P[1] = 3pq^2$$
$$P[2] = 3p^2q, \quad P[3] = p^3$$

Table B.3   Outcome of Three Coin Flips

Event	Outcome	Sum	Probability
1	000	0	qqq
2	001	1	qqp
3	010	1	qpq
4	011	2	qpp
4	100	1	pqq
6	101	2	pqp
7	110	2	ppq
8	111	3	ppp

It is easy to show that P[0] + P[1] + P[2] + P[3] = 1 as required. Notice the coefficient sequence [1 3 3 1] in the results above. This is recognized as the third line of Pasqual's triangle. The general result that, for n trials, the probability the sum will equal k, is given by the binomial distribution

$$P_n\left[sum = k\right] = c_k^n p^k \left(1 - p\right)^{n-k}$$
$$c_k^n = n!\big/k!(n - k)!$$

By using Sterling's formula

$$n! = \sqrt{2\pi}n^{n+.5}e^{-n}$$

after a lot of manipulation we can write

$$P_n\left[sum = k\right] = \exp\left[-(k - np)^2\big/2npq\right]\big/\sqrt{2\pi npq}$$

where we note that $np = \mu$, and $\sigma^2 = npq$ are the mean and variance of the original binomial distribution.

The two-dimensional GRV is given by the PDF

$$f_{x,y}\left(\alpha,\beta\right) = \frac{1}{2\pi\sqrt{1-\rho^2}} \exp\left[-\frac{\alpha^2 - 2\rho\alpha\beta + \beta^2}{2\left(1-\rho^2\right)}\right]$$

where we have assumed that the mean of each variable is zero and the variance of each is unity, GRV[0, 1]. For a general GRV[μ, σ], the PDF can be obtained by the transformation to a new variable α', $= (\alpha' - \mu)/\sigma$, (which is a good exercise for the reader)

$$\rho = \int_{-\infty}^{\infty}\int_{-\infty}^{\infty} \alpha\beta f_{x,y}\left(\alpha,\beta\right)d\alpha d\beta$$
$$-1 \le \rho \le 1$$

The symbol ρ is called the correlation coefficient. Finally, we have the two relations,

$$p_x\left(\alpha\right) = \exp\left(-\alpha^2\big/2\right)\big/\sqrt{2\pi}$$

independent of ρ, and

$$p_x\left(\alpha|\beta = \varepsilon\right) = \exp\left[-(\alpha - \rho\varepsilon)^2\big/2\left(1-\rho^2\right)\right]\big/\sqrt{2\pi\left(1-\rho^2\right)}$$
$$= GRV\left[\rho\varepsilon, \sqrt{\left(1-\rho^2\right)}\right]$$

This last expression illustrates an important point. If the correlation coefficient is $\rho = 1$ or -1, then there is no mystery as to what á would be, specifically $\alpha = \pm\varepsilon$. Furthermore, the uncertainty of this knowledge is zero (i.e., the variance is zero). This is reflected in the equation. Now suppose $\rho = 0$; then there is no help and we have the statistics on x as GRV[0, 1], which is the PDF of x without regard to y, as expected.

B.2.2   The Uniform Distribution

A second PDF that commonly appears is the uniform PDF. In this case any value of the random variable x is equally likely. Consider the child's toy of an arrow mounted on a piece of cardboard. The arrow is free to spin. The PDF of the angle θ is simply

$$p_\theta(\alpha) = 1/2\pi \quad 0 \le \alpha \le 2\pi$$

Question: You spin the arrow. What is the probability that the resulting angle is 33.11 degrees?

Answer: The probability is 0. This is one of the peculiarities of random variables, which can have a continuous output: an event of probability 0 that can happen.

Selected Bibliography

Davenport, W., Jr., *Probability and Random Processes: An Introduction for Scientists and Engineers*, New York: McGraw-Hill, 1970.

Wozencraft, J. M., and I. Jacobs, *Principles of Communication Engineering*, New York: John Wiley and Sons, 1967.

APPENDIX C
Error Correcting Codes

Error correcting codes (ECC) are a powerful method for increasing the performance of a communications system in AWGN. The basic idea is to add redundancy to a block of bits in such a way that errors that occur in transmission can be corrected giving an overall system performance. In this appendix, we will detail the two most common code types: block codes and convolutional codes. Additional FEC code structures are trellis codes, and low-density parity check codes.

This appendix is designed as a brief introduction to FEC, to familiarize the reader with some its basic concepts. More in depth presentations can be found in the references at the end.

C.1 Block Codes

As the name implies, block codes deal with a finite group of input bits at a time. There are three basic numbers associated with such codes:

1. The number of input bits in the block: k;
2. The number of output bits in the block: n;
3. The number of errors that can be corrected: t.

The general notation is simply a $[n, k, t]$ code. For example, the Golay code has the notation [23, 12, 3]. The rate of the code, r, is given by $r = k/n$.

The algebra of coding theory is Boolean logic. The multiplication operator $\otimes$ is the common AND gate and has the algebra given in Table C.1.

In the same manner, the addition operation $\oplus$ is the XOR having the truth table of Table C.2.

C.1.1 Generating the Codes

The objective is to map a block of k bits into another block of n bits. A common way to do this is to use matrix algebra. If we let the input bits be denoted by a k-bit column vector $[u]_k$, and the corresponding n-bit output by a column vector $[v]_n$, we can write

$$[v]_n = [G]_{n,k} [u]_k$$

where $[G]_{n,k}$ is called the generator matrix. Usually the generator matrix is of the form

Table C.1 Boolean Multiplication (AND) Logic

Input 1	Input 2	Output
0	0	0
0	1	0
1	0	0
1	1	1

Table C.2 Boolean Addition (XOR) Logic

Input 1	Input 2	Output
0	0	0
0	1	0
1	0	1
1	1	1

$$[G]_{n,k} = \begin{bmatrix} I_k \\ \cdots \\ P_{(n-k),k} \end{bmatrix}$$

where I_k is the $k \times k$ identity matrix, and $P_{(n-k),k}$ is called the parity array matrix. As an example, take

$$[G]_{6,3} = \begin{bmatrix} 100 \\ 010 \\ 001 \\ 101 \\ 110 \\ 011 \end{bmatrix}$$

The code word for the input vector [111] is then given by

$$\begin{bmatrix} 1 \\ 1 \\ 1 \\ 0 \\ 0 \\ 0 \end{bmatrix} = \begin{bmatrix} 100 \\ 010 \\ 001 \\ 101 \\ 110 \\ 011 \end{bmatrix} \begin{bmatrix} 1 \\ 1 \\ 1 \end{bmatrix}$$

by applying the basic logic rules of Tables C.1 and C.2. Table C.3 shows the complete coding structure.

Table C.3 Coding of the [6, 3] Code

Input Code Word	Output Code Word
000	000000
100	100110
010	010011
110	110101
001	011101
101	101011
011	011110
111	111000

Note that the first 3 bits of the output code word are the original message bits. This type of code is called *systematic*.

From Table C.1 it can be seen that for any two output code words, the code bit pattern differs in three positions. If we assume a single error, then we should be able to make the proper correction. For example, if we receive the word [000001], then no other code word with one error can reach this pattern. The proper decode to [000000] is then possible. On the other hand, if we receive the code word with two errors [000011], then either of the true code words [000000] or [101011] with two errors could produce the same result.

Let [e] be an n-bit vector that represents the errors that were incurred during the transmission. Then the received code word [r] can be written as

$$[r] = [v] + [e]$$

The task at hand is, given [r], how do we choose or decode [v] and, from there, the original data vector [u]? In Chapter 5 we saw that given a set of possible signals, the optimum detector correlates the actual received signal with each of the members of the possible signal, and chooses the largest value as the result. The same idea applies here. What we could do is perform the vector dot products

$$D_1 = \sum_{m=0}^{N-1} v_{l,m} r_m \quad 0 \le 1 \le 2^k - 1$$

where $v_{l,m}$ is the mth bit of the lth possible code word, and choose the largest. Realize that because of the binary operations, there could be a tie that must be resolved. To be specific, for the [23, 12, 3] Golay code, we would have to perform $2^{12} = 4,096$ 23-bit correlations. In general, the computation rate R_c required for real-time decoding is related to the information bit rate R by

$$R_c = \text{\# of operations / word time}$$
$$= \left(n2^k\right)/kT_b$$
$$= n2^k R/k$$

For the Golay code, we obtain $R_c = 276R$. Thus, the maximum rate R that can be decoded in real time requires a computer operation 276 times faster.

With modern data computing power, reasonable throughput rates are quite feasible. But when the development of FEC began, this power was not there by orders of magnitude, so the computational time was prohibitive. Thus, researchers in this field developed very sophisticated mathematical systems now known as algebraic coding theory.

To proceed, we now define the parity check matrix:

$$[H] = \begin{bmatrix} I_{n-k} \\ \cdots \\ P^T \end{bmatrix}$$

The useful property of $[H]$ is

$$[B] = [H]^T [G] = [0]$$

If we multiply both sides of the above equation by a message word $[u]$, we obtain

$$[B][u] = [H]^T [G][u] = [H]^T [v] = [0]$$

Now if the corrupted received vector is $[r] = [v] + [e]$, then the operation

$$[S] = [H^T][r] = [H^T]\{[v] + [e]\}$$
$$= [H^T][e]$$

yields an $n-k$ vector, $[S]$, called the syndrome, which is a function of the error pattern *only*. The last relation is a set of $n-k$ equations in the n possible errors. This is an underdetermined system so it is not possible to solve for all of the errors as we noted before. The whole of algebraic theory basically reduces to clever algorithms for efficiently recovering the errors given $[S]$.

There is a useful vector space interpretation of $[S]$. The output code word $[v]$ defines an n-dimensional vector space $[\hat{V}]$ with 2^n elements. The set of input code words generates a 2^k-dimensional subspace of $[\hat{V}]$ denoted by $[\hat{U}]$. This leaves a second subspace, called the complimentary space of dimension 2^{n-k}. The received code word $[r]$ can be anywhere in $[\hat{V}]$. From the above relations, it can be seen that the syndrome $[S]$ is that component of $[r]$ that is orthogonal to the code word subspace $[\hat{U}]$. Figure C.1 shows this interpretation.

We now consider what is gained by all of this encoding and decoding. The following development is designed to show the dominant behavior of the coding. To start, let $P_c = 1 - P_e$ be the probability that one of the n encoded bits (the common terminology is to call these bits "chips") is decoded correctly. Assuming that the decision on one chip is independent of the decision on another, the probability $P = 1 - P$, that the whole code word is decoded correctly is simply $P = P_c^n$. Note that P is related

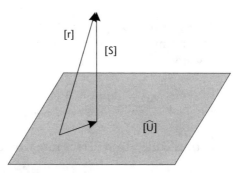

Figure C.1   The syndrome vector is the component of the received vector that is orthogonal to the subspace spanned by the code words.

to but not the same as the final individual bit error rate. For a code the corrects exactly t errors, but no more, the word error probability

$$P = \sum_{m=t+1}^{n} \text{Probability } (m \text{ chips in error})$$

In the limit of high E_c/N_0 (energy per encoded chip), the first term in the sum dominates, so we can write

$$P \approx \text{Probability } (t+1 \text{ chips in error})$$

Assuming that each chip in the code word is decoded independently, we have

$$\text{Probability } (t+1 \text{ chips in error}) = C_{t+1}^{n} P_e^{t+1} \left(1 - P_e^{n-t-1} \right)$$
$$\approx C_{t+1}^{n} P_e^{t+1}$$

where C_{t+1}^{n} is the binomial coefficient and again for large E_c/N_0. From Chapter 5, for binary transmission, the best expression for the chip error rate is

$$P_e = Q\left(\sqrt{2E_c/N_0} \right)$$
$$Q(x) = \frac{1}{2\pi} \int_{x}^{\infty} e^{-z^2/2} \, dz$$
$$\approx \frac{1}{z\sqrt{2\pi}} e^{-z^2/2}$$

for large z. The chip energy/bit is related to the bit energy/bit by the relation $nE_c = kE_b$, since we are transmitting n chips in the same time as we would transmit k bits. Combining all of these relations yields

$$P \approx C_k^{n} \left\{ \frac{1}{\sqrt{4\pi r E_b/N_0}} e^{-rN_b/N_0} \right\}^{t+1}$$

The bit error rate, P_b, is proportional to the word error rate. What we are after is an order of magnitude estimate that is controlled by the exponential term in the formula above, so we write

$$P_b \approx \left[e^{-rE_b/N_0} \right]^{t+1}$$

Now if we did not encode, the bit error rate would be in similar fashion

$$P_b \approx e^{-E_b/N_0}$$

First, note the probability of error for a chip is greater than if we just sent the bit. This is because the encoded chip is shorter in duration, so $E_c < E_b$. But the error correcting capability produces a net gain as long as a figure of merit

$$\beta = r(t+1) = k(t+1)/n > 1$$

For the [6, 3, 1] code we considered above, we have $\beta = 1$. This code provides no real gain for all of the effort to encode and decode.

We mention the popular set of block codes known as the Bose-Chaudhuri-Hocquenghem (BCH) codes. These codes have a general structure, for an integer m,

$$n = 2^m - 1$$
$$n - k \leq mt$$

Table C.4 lists BCH codes for $m = 5$. In that table we have calculated the code rate r, and also β.

As can be seen, β is smaller for low and high values of r, and tends to peak near $r = 0.5$. We can show this for BCH codes as follows. Take the second equation defining the code with the equals sign and substitute the expression for â. The result is

$$\beta = k(n-k)/(mn)$$

Now for fixed m, and hence n, differentiate with respect to k to find $k = n/2$ maximizes the expression, which in turn gives $r_{max} = 1/2$.

Table C.4   BCH Codes for $m = 5$

n	k	T	r	β
31	26	1	0.839	1.68
31	21	2	0.677	2.03
31	16	3	0.516	2.06
31	11	5	0.355	2.12
31	6	7	0.195	1.54

C.2   Convolutional Codes

The coding structure for convolutional codes is entirely different than that for block codes. The basic coding structure is shown in Figure C.2.

The input data enters into the set of three shift registers from the left. After every shift we compute 2 bits, Y_1 and Y_2, according to the logic shown. The mathematical operation is the XOR. Thus, for every 1 bit in, there are 2 bits out, producing a rate 1/2 code. The two individual data streams are then merged together. The number of registers—three in this case—is called the constraint length $K = 3$. The codes can be generalized with larger values of K (7 to 11 in most cases) and by adding additional logic to produce more output bits Y_3, Y_4, and so forth. The code notation is obtained from the particular combinations of those delay elements connected to the XOR. In this example the output Y_1 can be represented by the 3-bit vector [1, 1, 1]. Converting this string to octal notation, we have 7_8. Consequently, this code is denoted by $[5, 7]_8$.

To see the encoder operation (Table C.5), we start with an input code

$$[u] = [1,1,0,1,1,\ldots]$$

and calculate step by step the output code word [v].

From the table we see that the encoded data stream is

$$[v] = [1,1,0,1,1,0,0,\ldots]$$

It is *very* instructive to consider the coding of two different input sequences $A = [0, 0, 0, m_1, m_2, \ldots]$, and $B = [1, 0, 0, m_1, m_2, \ldots]$ for the [5, 7] code discussed so far. Here, m_1 and m_2 denote arbitrary bits. Table C.6 shows the states of the three shift registers as the coding progresses.

We now make an extremely important observation. No matter what the next bit is in either message, the output bits will be the same for the same input bit, m. This is shown in Table C.7.

What this means is that after the first 6 received bits (2 out for 1 in), we can merge the correlation processes. The largest correlation up to this point would be the survivor. There are other bit combinations that produce this type of result, both at the start of the message and further down. The correlation decision process can then be vastly simplified.

This result has been formalized as follows. The concept is a finite state machine (FSM). The last $k - 1$ registers define 2^{k-1} states. In the case at hand there are four states generally labeled, a = 00, b = 10, c = 01, and d = 11. We start by loading in two

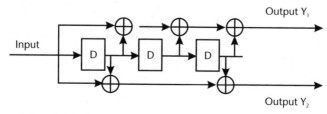

Figure C.2   Convolutional encoder.

Table C.5   Convolutional Encoder Operation

Time Step	Reg #1 (input bit)	Reg #2	Reg #3	Bit Y_1	Bit Y_2
1	1	0	0	1	1
2	1	1	0	0	1
3	0	1	1	1	0
4	1	0	1	0	1
5	1	1	0	0	0

Table C.6   Encoding Different Input Bit Patterns *A* and *B*

Reg A1	Reg A2	Reg A3		Reg B1	Reg B2	Reg B3
0	0	0		1	0	0
0	0	0		0	1	0
0	0	0		0	0	1

Table C.7   Encoding Same Input Bit Patterns *A* and *B*

Reg A1	Reg A2	Reg A3		Reg B1	Reg B2	Reg B3
m_2	0	0		m_2	0	0

zero (state a). The first bit is placed in the first register. We now clock the system producing 2 output bits according to Table C.5, and depending on whether the first bit is a 1 or a 0. Note that the first information bit has moved into the second register and the state is [b0]. We continue this process to obtain the state machine transition matrix as shown in Table C.8.

The most useful presentation of this concept is the trellis shown in Figure C.3.

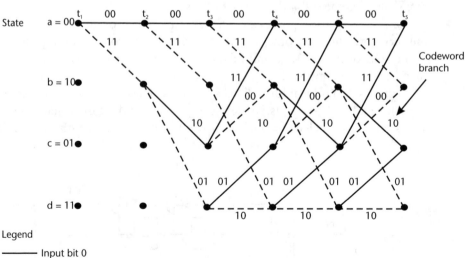

Figure C.3   Encoding state space trellis for the [3, 5], *k* = 3 convolutional code.

Table C.8 Coder Transition Matrix

Current Input State	Input Input Bit	Final Output State
00 (a)	0	00(a)
00(a)	1	10(b)
01(c)	0	00(a)
01(c)	1	10(b)
10(b)	0	01(c)
10(b)	1	11(d)
11(d)	0	01(c)
11(d)	1	11(d)

On each transition from one state to the next, Figure C.3 shows the 2 bits of encoded data. For instance, a transition from state [a] to state [b] always produces the 2-bit pattern [1, 1]. From this figure we also verify that the two t sequences [A] and [B] merge after the third input bit.

C.2.1 Punctured Codes

The example discussed so far produces convolutional codes with rate r = 1/n, where n is the number of combinatorial circuits [A, B, ...] that we choose to add. But what about codes like r = 2/3? Codes of r = k/n (generally) are easy to obtain from block codes, as shown in Table C.4. In Figure C.4 we show a convolutional coding structure that produces an r = 2/3 code. In this case we shift in 2 bits at a time and perform the indicated operations, then take 3 bits out. In the figure we see two groups of two shift registers.

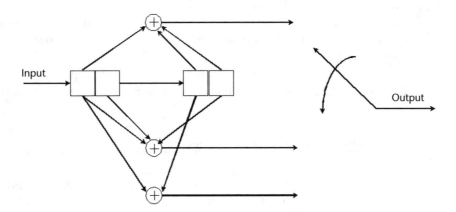

k = 2 input bits; n = 3 output bits; L = constraint length = 2 groups of k bits

Figure C.4 Rate 2/3, K = 2 encoder.

The constraint length is, by convention, $K = 2$, not $K = 4$.

The coding and decoding trellis states for this code can be generated in the same manner as before. The only problem is that these structures become much more complex. The neat solution to this problem, which is commonly used, is to take a code like $r = 1/2$, as shown above, and throw away 1 out of every 4 encoded bits. This produces a rate 2/3 code; this process is called *puncturing*. Which of the 4 bits eliminated is determined by computer simulations. Table C.9 shows a particular set of punctured codes for the case where the rate is of the form, $r = n/(n + 1)$. The numbers in ( ) are the octal notation of the encoder connections as just described.

C.2.2 Decoding Convolutional Codes

Just like the block codes, the optimum decoder would correlate the received code word with all possible code words. The one with the closest match would be taken as the message. The Viterbi decoding algorithm implements this concept by taking advantage of the coding trellis of Figure C.3. This idea is presented in Figure C.5.

In Figure C.5, the top row is the input sequence from the previous coding example, and the second row is the corresponding output sequence as previously noted in Table C. The third row represents the decoded bits of the received signal, errors and all. Refer back to Figure C.3 along with Figure C.5. Now start at time t_1. From Figure C.3 we see that that the [a] to [a] transition with an input bit of [1] produces the output bits [0, 0]. But the received sequence [Z] at this point is [1, 1]. There are two disagreements as noted in Figure C.5. This is known as the branch metric. Again back to Figure C.3—we see that the output bit, for an input bit of [1], is the bit pair [1, 1]. There are no disagreements with the received bit pattern as noted on Figure C.5.

The decoding is simple. Proceed from node to node and determine the corresponding branch metrics. Starting from t_1, there is only one correct path through the trellis as established in the encoding operation. For each path we keep a running total of the appropriate branch metrics, called the path metric. When two paths merge, the path metric with the smallest (not the largest in this normalization) is kept, and the other is discarded. If there is a tie, then one must flip a coin to choose one path and continue. A wrong guess produces errors. This process continues to the end of the message. Finally, the surviving path with the lowest metric is selected for output.

Unlike block codes, convolutional codes do not have a specific length. In theory, at least, the encoded data stream could go on forever. At the receiver, this becomes a

Table C.9 Punctured Code Patterns

Code Rate	3	4	5	6	7	8	9
1/2	1 (5)	1 (15)	1 (23)	1 (53)	1 (133)	1 (247)	1 (561)
	1 (7)	1 (17)	1 (75)	1 (75)	1 (171)	1 (371)	1 (753)
2/3	10	11	11	10	11	10	11
	11	10	10	11	10	11	10
3/4	101	110	101	100	110	110	111
	110	101	110	111	101	101	100

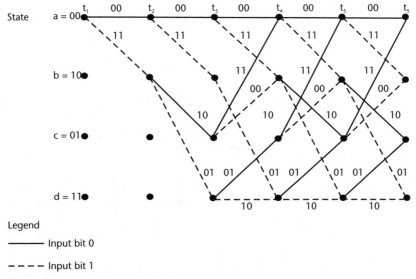

Figure C.5   Decoding trellis for the [3, 5], $k = 3$ convolutional code.

problem, of course, since one never reaches the end of the message so decoding can begin. In many practical cases associated with the wireless industry, the codes are terminated. For cell phone operation, the analog voice is converted into digits by a device called a vocoder that operate on blocks of data, called frames, roughly 5 to 20 msec long. The other possibility is to artificially truncate the path metric evaluation at some point and select the best match as the survivor. Notice that if the wrong path is chosen, there can be a large number of errors. The result of this is that the error pattern in convolutional codes tends to arrive in bursts as shown in Figure C.6.

C.2.3   Hard Versus Soft Decoding

Note that in both the block and convolutional decoders, a decision was made as to whether an encoded chip was a 1 or a 0 *before* the decoder. This operation is called

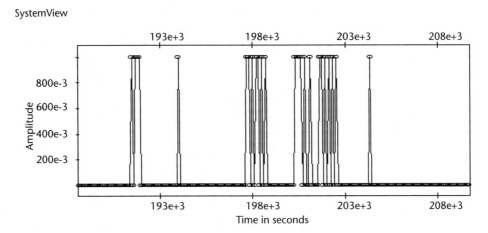

Figure C.6   Burst nature of convolutional code error patterns.

hard decoding; computers like 1s and 0s. The ideal decoder would recover the chips one at a time. We can write then

$$r_k = d_k + n_k$$

As discussed in Chapter 5, the optimum detector would perform the correlations

$$D_p = \sum_{m=0}^{N-1} v_{p,m} r_m \quad 0 \le p \le 2^k - 1$$

and choose the largest as the message. This is called a soft decision decoder. By making hard decisions, the decoder is suboptimum and there is a theoretical loss of 2.5 dB.

The problem with block codes is that it is difficult to modify them for soft decision decoding. The same is not true for convolutional codes. But as we have emphasized many times here, some quantization is required for computer operations. In other words, how many bits should we use to represent [r]? The established answer from simulations and theory is that 3 bits of quantization produces a decoder that recovers 2 of the 2.5 dB of loss of the hard decoder.

Selected Bibliography

Shu, L., and D. J. Costello, *Error Control Coding, Fundamentals and Applications*, Upper Saddle River, NJ: Prentice Hall, 1983.

Sklar, B., *Digital Communications Fundamentals and Applications*, Upper Saddle River, NJ: Prentice Hall, 2001.

Trivia Question

Trivia Question: Who coholds the earliest patent awarded in spread spectrum technology?

Amazing Answer: The Hollywood actress Hedy Lamarr. She is probably best known for her role as Delilah in the 1949 movie *Samson and Delilah*.

With colleague George Antheil, she discussed radio guided torpedoes. The problem was how to keep the radio link from being jammed. She suggested the idea of frequency hopping, as described in Chapter 6. Antheil provided the synchronization concept that kept the transmitter and receiver hopping in unison, they were awarded UP Pat. No. 2292387, on August 1, 1942.

Selected Bibliography

http://www.inventions.org/culture/female/lamarr.html.

About the Author

Maurice Schiff received his Ph.D. from the University of California, San Diego in 1969. He has more than 35 years of experience in the field of digital communications, spread spectrum systems, digital signal processing, and radar systems. From 1973 to 1978 Dr. Schiff was at ITT, Ft. Wayne, Indiana, where he developed the Frequency Hopping architecture for the USA Single Channel Air Ground Radio System (SINCGARS). He coholds the patent for the synchronization system. From 1992 to 2003 Dr. Schiff was the CTO of Elanix, Inc., where he was involved with the development of the SystemVue PC-based simulation tool. Currently he is with ITT/ACD in Westlake Village, California, developing algorithms for GPS systems. Dr. Schiff has given short seminars in communication theory and spread spectrum systems at various trade shows. He has also produced five tutorial articles related to spread spectrum systems, FFT processing, and baseband pulse shaping for the Web site Techonline (http://www.tolu.com).

Index

Recent Titles in the Artech House Mobile Communications Series

John Walker, Series Editor

3G CDMA2000 Wireless System Engineering, Samuel C. Yang

3G Multimedia Network Services, Accounting, and User Profiles, Freddy Ghys, Marcel Mampaey, Michel Smouts, and Arto Vaaraniemi

802.11 WLANs and IP Networking: Security, QoS, and Mobility, Anand R. Prasad, Neeli R. Prasad

Advances in 3G Enhanced Technologies for Wireless Communications, Jiangzhou Wang and Tung-Sang Ng, editors

Advances in Mobile Information Systems, John Walker, editor

Advances in Mobile Radio Access Networks, Y. Jay Guo

Applied Satellite Navigation Using GPS, GALILEO, and Augmentation Systems, Ramjee Prasad and Marina Ruggieri

CDMA for Wireless Personal Communications, Ramjee Prasad

CDMA Mobile Radio Design, John B. Groe and Lawrence E. Larson

CDMA RF System Engineering, Samuel C. Yang

CDMA Systems Capacity Engineering, Kiseon Kim and Insoo Koo

CDMA Systems Engineering Handbook, Jhong S. Lee and Leonard E. Miller

Cell Planning for Wireless Communications, Manuel F. Cátedra and Jesús Pérez-Arriaga

Cellular Communications: Worldwide Market Development, Garry A. Garrard

Cellular Mobile Systems Engineering, Saleh Faruque

The Complete Wireless Communications Professional: A Guide for Engineers and Managers, William Webb

EDGE for Mobile Internet, Emmanuel Seurre, Patrick Savelli, and Pierre-Jean Pietri

Emerging Public Safety Wireless Communication Systems, Robert I. Desourdis, Jr., et al.

The Future of Wireless Communications, William Webb

GPRS for Mobile Internet, Emmanuel Seurre, Patrick Savelli, and Pierre-Jean Pietri

GPRS: Gateway to Third Generation Mobile Networks, Gunnar Heine and Holger Sagkob

Universal Wireless Personal Communications, Ramjee Prasad

WCDMA: Towards IP Mobility and Mobile Internet, Tero Ojanperä and
  Ramjee Prasad, editors

Wireless Communications in Developing Countries: Cellular and Satellite Systems,
  Rachael E. Schwartz

Wireless Intelligent Networking, Gerry Christensen, Paul G. Florack, and
  Robert Duncan

Wireless LAN Standards and Applications, Asunción Santamaría and
  Francisco J. López-Hernández, editors

Wireless Technician's Handbook, Second Edition, Andrew Miceli

For further information on these and other Artech House titles, including previously
considered out-of-print books now available through our In-Print-Forever® (IPF®)
program, contact:

Artech House   Artech House
685 Canton Street   46 Gillingham Street
Norwood, MA 02062   London SW1V 1AH UK
Phone: 781-769-9750   Phone: +44 (0)20 7596-8750
Fax: 781-769-6334   Fax: +44 (0)20 7630-0166
e-mail: artech@artechhouse.com   e-mail: artech-uk@artechhouse.com

Find us on the World Wide Web at: www.artechhouse.com

CHECK FOR ___!___ PARTS

(1 CO)